T0136404

# An Introduction to Virtual Sound Barriers

# An Introduction to Virtual Sound Barriers

Xiaojun Qiu

**CRC Press**
Taylor & Francis Group
Boca Raton  London  New York

CRC Press is an imprint of the
Taylor & Francis Group, an **informa** business

CRC Press
Taylor & Francis Group
6000 Broken Sound Parkway NW, Suite 300
Boca Raton, FL 3487-2742

First issued in paperback 2021

© 2020 by Taylor & Francis Group, LLC
CRC Press is an imprint of Taylor & Francis Group, an Informa business

No claim to original U.S. Government works

ISBN-13: 978-1-03-209156-3 (pbk)
ISBN-13: 978-0-8153-4810-8 (hbk)

This book contains information obtained from authentic and highly regarded sources. Reasonable efforts have been made to publish reliable data and information, but the author and publisher cannot assume responsibility for the validity of all materials or the consequences of their use. The authors and publishers have attempted to trace the copyright holders of all material reproduced in this publication and apologize to copyright holders if permission to publish in this form has not been obtained. If any copyright material has not been acknowledged please write and let us know so we may rectify in any future reprint.

Except as permitted under U.S. Copyright Law, no part of this book may be reprinted, reproduced, transmitted, or utilized in any form by any electronic, mechanical, or other means, now known or hereafter invented, including photocopying, microfilming, and recording, or in any information storage or retrieval system, without written permission from the publishers.

For permission to photocopy or use material electronically from this work, please access www.copyright.com (http://www.copyright.com/) or contact the Copyright Clearance Center, Inc. (CCC), 222 Rosewood Drive, Danvers, MA 01923, 978-750-8400. CCC is a not-for-profit organization that provides licenses and registration for a variety of users. For organizations that have been granted a photocopy license by the CCC, a separate system of payment has been arranged.

**Trademark Notice:** Product or corporate names may be trademarks or registered trademarks, and are used only for identification and explanation without intent to infringe.

**Publisher's Note**
The publisher has gone to great lengths to ensure the quality of this reprint but points out that some imperfections in the original copies may be apparent.

**Visit the Taylor & Francis Web site at**
**http://www.taylorandfrancis.com**

**and the CRC Press Web site at**
**http://www.crcpress.com**

# Contents

# *Preface*

Active noise control is a method of reducing existing noise via the introduction of controllable secondary sources that affect radiation, transmission, and reception of the original primary noise source. It can provide better solutions to low-frequency noise problems than passive noise control methods when there are weight and volume constraints. The fundamental theories and methods of active noise control have been well established over the past four decades, and successful applications include active noise control systems in headsets, propeller aircraft, and cars. Virtual sound barriers are a special form of active noise control systems for applications where natural air ventilation, light, and easy access are desired for sound barriers.

A virtual sound barrier is an active noise control system that uses arrays of loudspeakers and microphones to create a quiet zone of a useful size in a noisy environment, just like an ordinary sound barrier but without much blocking of air, light, and access. This system can be used to reduce sound propagation, radiation, and transmission from noise sources, or to reduce noise levels around a few persons in a noisy environment. In the past 10 years, many researchers have contributed to this topic and some typical applications have emerged. The objective of this book is to bring together the results of contemporary research relating to this topic and present them in a systematic way.

The first chapter introduces the concepts in sound propagation and the principles of various sound barriers that are used for sound propagation control. The basic wave phenomena relating to sound propagation, such as acoustic reflection, absorption, scattering, and diffraction, are introduced first, and then the corresponding parameters such as reflection coefficient, absorption coefficient, scattering coefficient, transmission loss, and insertion loss are explained. The principles of passive sound barriers and active sound barriers are discussed next as well as the methods used for their design. Finally, the history, principles, and design methods of virtual sound barriers are introduced. The contents are focused on the fundamental aspects of sound propagation and barriers to serve as an introduction to students and engineers with a basic knowledge in mathematics and physics.

The second chapter focuses on planar virtual sound barriers. The problem is defined first, so that the concept of planar virtual sound barriers can be introduced. The use of planar virtual sound barriers is then applied to the control of plane wave propagation in free fields, the control of plane wave propagation through a finite aperture, the control of sound radiation from an opening of an enclosure, and the control of sound transmission via an opening into an enclosure. Both the boundary control and surface control systems are discussed, and their upper-limit frequencies for effective control are considered. The methods for designing planar virtual sound barriers for these applications are given and the mechanisms for the control are explored.

The third chapter introduces three-dimensional virtual sound barriers, which can be used to create a quiet zone in a noisy environment just like a passive sound barrier, but without much blocking of air, light, and access. The effects of a diffracting sphere inside the quiet zone, a reflective surface near the system, and the cost functions for optimizing the system for better performance are discussed. The virtual sensor algorithms developed for local active noise control systems are reviewed and the effects of virtual sensor locations on system performance are discussed. It is found that system performance can be improved significantly with the use of virtual sensors when the distances between the secondary sources and the physical sensors are small.

Virtual sound barrier systems can be implemented in situations where both sound control and free air ventilation are required. The fourth chapter introduces two examples of virtual sound barrier applications for noise control. The first is the control of noise radiation by power transformers in a hemi-closed space with a planar virtual sound barrier, where an array of loudspeakers is installed evenly at the open door of a room to reduce the noise radiation generated by the transformers inside. The second is the control of sound transmission into a room through an open window, where an array of loudspeakers is installed at an open window of a room to reduce the transmission of noise generated by outside traffic into the room. Finally, the implementation issues relating to these applications are discussed.

Although the existing research has demonstrated that it is feasible to develop virtual sound barrier systems, there are still obstacles to their practical use, due to the narrow bandwidth and low-frequency range that can be used for effective control, the complexity, and the high cost of the systems. The last chapter summarizes the research progress made concerning virtual sound barriers, lists the problems and challenges, and gives the perspectives. If the upper-limit frequency of virtual sound systems can be increased, they have the potential to be applied to many sound control scenarios with ventilation and/or access requirements.

More than 130 references are provided at the end of the book. The list might not be completely comprehensive, but it provides readers with a good starting point for further study. It is my hope that this book can provide a clear picture of the current status of research on virtual sound barriers, so that researchers and engineers can be tempted to carry out the research and development projects relating to this topic in order to broaden the applications of virtual sound barriers.

Although I take full responsibility for any shortcomings of this book, I would like to thank some of my former postgraduate students whose research work forms the basis of the book. In particular, I would like to thank Dr. Haishan Zou, who explored three-dimensional virtual sound barriers and applications of active and virtual sound barriers; Dr. Jiancheng Tao, who applied planar virtual sound barriers on transformer noise control; Dr. Ning Han, who investigated sound intensity control for active sound barriers and applied virtual sound barriers for scattering sound control; Dr. Weisong Chen, who developed compound secondary sources for active sound barriers; Dr. Huahua Huang, who investigated active noise control on naturally ventilated windows; Dr. Sipei Zhao, who developed a new calculation method for calculating sound diffraction around barriers and applied virtual sound barriers for outdoor traffic noise reduction; Dr. Shuping Wang, who used planar virtual sound barriers to control sound radiation from a cavity and developed boundary control systems; and Mr. Kang Wang, who applied planar virtual sound barriers to control sound transmission into a room and discovered the upper-limit frequency of the systems. Some of the research was supported by the National Science Foundation of China (Projects 10304008, 10674068, and 11474163) and under the Australian Research Council's Linkage Projects funding scheme (LP140100987).

I would like to thank my former mentors, Professors Jiazheng Sha and Colin H. Hansen, for their supervision, trust, and support since I entered the field of active noise control research more than 20 years ago. Many thanks to my colleagues, Professors Jie Pan and Jianchun Cheng, for their collaboration, encouragement, and criticism during the research. I am grateful to my family, particularly my wife Donna and son Harry, for their understanding and support.

**Xiaojun Qiu**

# *Author*

 **Xiaojun Qiu, PhD,** is professor of audio, acoustics, and vibration at the Center for Audio, Acoustics, and Vibration in the Faculty of Engineering and Information Technology, University of Technology Sydney, Australia. He earned his Bachelor's and Master's degrees from Peking University in 1989 and 1992, respectively, and his PhD from Nanjing University in 1995, all majoring in Acoustics. He worked at the University of Adelaide, Australia, as a research fellow in the field of active noise control from 1997 to 2002, at the Institute of Acoustics of Nanjing University, China, as professor of acoustics and signal processing from 2002 to 2013, and at RMIT University, Melbourne, Australia, as professor of design in audio engineering from 2013 to 2016. Dr. Qiu joined the University of Technology Sydney in 2016. His main research areas include noise and vibration control, room acoustics, electro-acoustics, and audio signal processing, particularly with regard to applications of active control technologies. He is a fellow of the Audio Engineering Society, and serves as an associate editor for the *International Journal of Acoustics and Vibration*, and as an associate technical editor for the *Journal of Audio Engineering Society*. He has published three books, five book chapters and more than 400 technical papers, as well as being the principal investigator for numerous projects. He has also been granted more than 50 patents on audio acoustics and audio signal processing.

# 1

## Introduction

## 1.1 Sound Propagation

Sound is a longitudinal mechanical wave, where the displacement of the medium at each point is normal to the local wave-front surface when the disturbance is traveling in a medium (Morfey, 2001). Sound speed is different in different mediums, in air under normal atmosphere pressure at 20°C, the sound speed is approximately 343 m/s. There is usually an energy loss when sound propagates in a medium. For example, when a sound wave propagates in porous materials the porous gas-filled medium can be treated as an equivalent uniform medium for analysis purposes. So a propagation factor $e^{j\omega t - \gamma x}$ can be used to describe the dependence of the propagating wave on time $t$ and the propagation coordinate $x$ (Bies, Hansen, and Howard, 2018). Here, $\omega = 2\pi f$ is the angular frequency and $f$ is the wave frequency. The *propagation constant* $\gamma$ is also called the *propagation coefficient*, which is a complex number that can be represented by,

$$\gamma = \sigma + jk \tag{1.1}$$

where the *propagation wave number* $k = \omega/c$ is also called the *phase coefficient*, $\sigma$ is called the *attenuation coefficient*, and $c$ is the *speed of sound* in the medium. Although the energy loss can be caused by some kinds of energy dissipation, such as absorption, the term "attenuation coefficient" is used in this book to describe the reduction in amplitude of a progressive wave with the distance in the propagation direction in the medium by $e^{-\sigma x}$. This coefficient is the amplitude attenuation coefficient instead of the energy attenuation coefficient.

The attenuation of a propagating sound caused by air can usually be neglected in the low-frequency range. For example, the sound pressure level attenuation is about 1 dB for a sound wave at 250 Hz traveling over 1000 m; however, for a sound wave at 4000 Hz, the attenuation can be from 24 dB to 67 dB depending on the temperature and humidity of the air (Bies, Hansen, and Howard, 2018). In the sound propagation prediction scheme, this sound absorption factor is usually considered separately, so it is not included in the discussion of the sound barriers in this book.

### 1.1.1 Sound Absorption and Absorption Coefficient

When a propagating sound wave encounters a different medium or space discontinuity in the media, reflection, diffraction, and/or transmission of waves occur, where the incident wave arriving at the boundary interacts with it to produce waves traveling away from the boundary (Morfey, 2001). The reflected wave or scattered waves follow certain rules. For example, for the specular reflection in which a plane incident wave is reflected by a uniform plane boundary, the normal wave number component of the incident field is reversed on reflection, and the wave number component parallel to the boundary is unaltered, so the angle of reflection is equal to the angle of incidence. Figure 1.1 summarizes the relationships of different kinds of energy when a propagating wave is incident upon a layer of porous material. The total input energy brought from the incident wave is $E_i$, which is equal to the summation of $E_r$, $E_s$, $E_a$ and $E_t$, i.e., the energy reflected and scattered from the boundary, the energy dissipated inside the porous material layer, and the energy transmitted through the layer.

There is usually an energy loss when a sound is reflected from a boundary, and the changes in amplitude and phase that take place during the reflection can be represented by the *complex reflection factor* or *sound pressure reflection coefficient* as

$$R = \frac{p_r}{p_i} \tag{1.2}$$

where $p_r$ and $p_i$ are the complex amplitudes of the reflected and incident waves, respectively. The sound pressure reflection coefficient is a property of

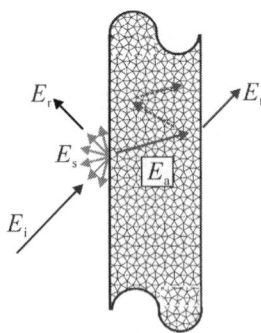

**FIGURE 1.1**
Sound reflection, scattering, absorption, and transmission when a propagating sound wave encounters a layer of porous material.

the boundary, and its magnitude and phase values depend on the frequency as well as the direction of the incident wave.

The acoustic *absorption coefficient* at a boundary, also called *sound absorption factor* or *sound power absorption coefficient*, is defined as the fraction of the incident acoustic power arriving at the boundary that is not reflected, and is therefore regarded as being absorbed by the boundary (Morfey, 2001),

$$\alpha = 1 - \frac{E_r}{E_i} \tag{1.3}$$

where $E_r$ and $E_i$ are the reflected acoustic power from and the incident acoustic power arriving at the boundary, respectively. The values of the acoustic absorption coefficient range from 0 to 1. A value of 0 refers to the condition of total reflection whereas 1 indicates perfect absorption without any reflection. It is also a function of frequency and incident wave direction. The (random incidence) *statistical absorption coefficient* can be obtained by,

$$\alpha_{st} = \frac{1}{\pi} \int_0^{2\pi} d\varphi \int_0^{\pi/2} \alpha(\theta) \cos\theta \sin\theta d\theta \tag{1.4}$$

where $\theta$ and $\varphi$ are the angle of elevation and azimuth, respectively. The *Sabine absorption coefficient* is the random incidence absorption coefficient deduced from the reverberation time measurement via the Sabine equation. Sound absorbers are usually employed for adjusting the reverberation in rooms, suppressing undesired sound reflections from remote walls or sound barriers, and reducing the acoustical energy density, and hence the sound pressure level, in noisy rooms or between parallel sound barriers. It is often difficult to obtain satisfying absorption at low frequencies with porous absorbers due to the required large thickness of the material.

### 1.1.2 Sound Insulation and Transmission Loss

When a sound wave is incident upon a partition, some of it is reflected back to the incidence medium and some is transmitted through the partition, as shown in Figure 1.1. The fraction of incident energy that is transmitted is called the *transmission coefficient*. If the scattering energy is small and can be neglected, the relationship between the transmission, reflection, and attenuation energy is

$$\tau = 1 - \gamma - \xi \tag{1.5}$$

where $\gamma$ and $\xi$ are the ratios of the sound power reflected back to the incident side and dissipated in the partition to the incident sound power, respectively.

The *transmission loss* is defined as the logarithm of the reciprocal of the transmission coefficient,

$$\text{TL} = -10\log_{10}\tau \tag{1.6}$$

In general, the transmission loss through a porous layer depends upon the angle of incidence and is a function of material density, thickness, flow resistivity, and frequency. In the low-frequency range, the transmission loss of common porous layers is usually less than 20 dB, but can be greater than 20 dB in the high-frequency range. The transmission loss usually increases with the material density, thickness, flow resistivity, and frequency. Passive sound barriers usually use woods, steel sheets, bricks, or concrete blocks, whose transmission loss is generally greater than 20 dB.

### 1.1.3 Sound Scattering and Scattering Coefficient

An obstacle or inhomogeneity in the path of a sound propagating path causes scattering if the secondary sound spreads out from it in a variety of directions, as shown in Figure 1.1 (Pierce, 1981). Although the scattered polar responses can give much information about the scattering, they yield too much detail, and a single value is desired in practice to allow easy comparison of diffuser quality. There is also a need for a *scattering coefficient* to evaluate the amount of dispersion generated by a surface to allow accurate predictions using geometric room acoustic models (Cox and D'Antonio, 2009). The geometric models use a scattering coefficient to determine the proportion of the reflected energy that is reflected in a specular manner and the proportion that is scattered.

The scattering coefficient $s_\theta$ for an incident wave arriving at the boundary with an angle $\theta$ is defined as the value calculated by one minus the ratio of the specularly reflected acoustic energy to the total reflected acoustic energy,

$$s_\theta = 1 - \frac{E_s}{E_r} \tag{1.7}$$

where $E_s$ is the specularly reflected acoustic energy and $E_r$ is the total reflected acoustic energy from the incident acoustic power arriving at the boundary. Theoretically, $s_\theta$ takes a value between 0 and 1, where 0 means a totally specularly reflecting surface and 1 means a totally scattering surface. Without the subscript, $s$ is the random incidence scattering coefficient, which is defined as the value calculated by one minus the ratio of the specularly reflected acoustic energy to the total acoustic energy reflected from a surface in a diffuse sound field. The scattering coefficient is a rough measure of the degree of the scattered sound. It is a simplified representation of the true reflection and scattering behaviors to characterize surface scattering for use in geometrical room modeling programs.

### 1.1.4 Sound Diffraction and Insertion Loss

Diffraction is non-specular reflection or scattering of sound waves by an object or boundary, particularly into the shadow zone, which is a region that rays emitted by a source cannot penetrate (Morfey, 2001). There is no sharp distinction between the two concepts of diffraction and scattering. When the scattering object is large compared with the wavelength of the scattered sound, the scattered sound is usually described as reflected or diffracted (Morse and Ingard, 1968). The physical mechanisms and effects are really the same, but the relative magnitudes differ enough for there to seem to be a qualitative difference. Behind the object there is a shadow or "dark" region, where the pressure amplitude is vanishingly small; at the front of the object is the "bright" region, where there is a combination of the incident wave and the wave reflected from the surface of the scattering object. At the edge of the shadow, the wave amplitude drops continuously from its value in the bright region to that in the dark region. For example, as shown in Figure 1.2, diffraction involves a change in the direction of the sound waves as they pass around a barrier in their path. The amount of diffraction (the sharpness of the bending) increases with wavelength. When the wavelength of the wave is much smaller than the obstacle or opening, no noticeable diffraction occurs, and reflection dominates.

A sound barrier placed between a noise source and a receiver can be used to reduce the direct sound observed by the receiver (Bies, Hansen, and Howard, 2018). In Figure 1.2, when a sound barrier is interposed, the dashed lines at $\theta = 3\pi - \theta_s$ and $\theta = \theta_s - \pi$ subdivide the sound field into three separated regions, I–III. When the receiver is in region I, the total sound field $p_p$ is composed of the directed sound by the primary source $p_i$, the reflected sound by the barrier $p_r$, and the diffracted sound by the edge of the barrier $p_d$. When the receiver is in region II, the total sound field is composed of the directed sound by the primary source $p_i$, and the diffracted sound by the edge of the

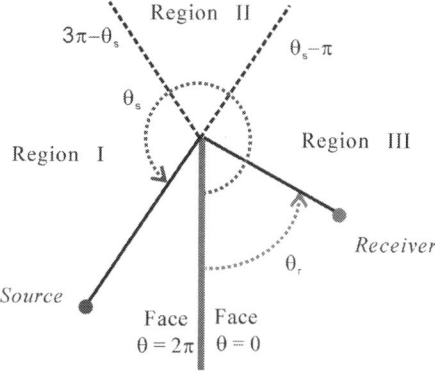

**FIGURE 1.2**
Three regions of sound diffraction by the top edge of a sound barrier.

barrier $p_d$. When the receiver is in region III, i.e., the "dark region", only the diffracted sound $p_d$ exists in the sound field (Guo and Pan, 1998b). The term, *insertion loss* (IL) is commonly used to describe the noise reduction performance of a sound barrier.

$$IL = 20\log_{10}\left|\frac{p_i(\mathbf{r})}{p_b(\mathbf{r})}\right| \tag{1.8}$$

where $p_i(\mathbf{r})$ and $p_b(\mathbf{r})$ are the sound pressure without and with the barrier at location $\mathbf{r}$, respectively.

## 1.2 Passive Sound Barriers

A sound barrier is any large structure that prevents line-of-sight sound transmission between a source and receiver, usually outdoors (Morfey, 2001). Effective blockage of the transmission path requires the structure to be acoustically opaque as well as large, compared with the sound wavelength. Barriers can be considered as a form of partial enclosure that is intended to reduce the direct sound field radiated in one direction only. For non-porous barriers having sufficient surface density, the sound reaching the receiver in the shadow is entirely due to the diffraction around the barrier boundaries. The diffraction sets the limit on the noise reduction that may be achieved, so the barrier surface density can be chosen to be just sufficient for the noise reduction at the receiver to be diffraction limited (Bies, Hansen, and Howard, 2018). For this purpose, the barrier surface density usually only needs to just exceed 20 kg/m².

Practically used sound barriers have different shapes, such as normal straight-edged barriers, top-bended barriers (T-shape, Y-shape, or half-Y-shape: Piacentini, Invernizzi, and Pannesse, 1996; Duhamel, 1996), inclined barriers (wave-trapping: Yang, Pan, and Cheng, 2013), and barriers with multiple edges (Min and Qiu, 2009). A thin rigid barrier is the fundamental form of all these sound barriers, and studies of wave diffraction by it have been a subject of interest for more than two centuries. A mathematically rigorous solution for the diffraction of a plane wave and a cylindrical wave incident on a half-plane was formulated by Sommerfeld (1896; Li and Wong, 2005), with MacDonald (1915) extending this approach to spherical incident waves. Hadden and Pierce (1981) applied this approach to the cases of diffracting wedges and wide-edged barriers. In these approaches, the space is artificially divided into three regions and the derivation process of the solutions is quite complicated.

Different from the wave-based methods, several methods based on geometrical theory have been used, mainly for calculation in high-frequency regions. Keller (1962) developed a geometrical theory of diffraction (GTD) by introducing diffracted rays in addition to the usual rays of geometrical optics and acoustics, in which various diffraction laws, analogous to the laws of reflection and refraction, were employed to characterize the diffracted rays. To overcome the singularity and caustics problem at and near the shadow boundaries of this method, the uniform GTD was developed, which utilized a uniform asymptotic method to eliminate the singularities at and near the shadow boundaries, but behaved similarly as the GTD method outside these regions (Pathak, Carluccio, and Albani, 2013).

Most of the abovementioned prediction schemes are based on the idea of a semi-infinite screen without the effect of ground reflection. The calculation of the sound field with a barrier on the ground is more complicated. The MacDonald solution was used to investigate the effect of the ground on the sound reduction of barriers (Jonasson, 1972), and the diffraction by a screen above an impedance boundary was studied with the "physical optics" method (Thomasson, 1978). Isei et al. (1980) compared five different diffraction theories with a model for ground impedance, and L'Esperance et al. (1989) combined the Hadden–Pierce method with the image source method to investigate the insertion loss of absorbent barriers on the ground. Most of these methods employed the image source or complex image source methods to deal with the finite impedance plane. de Lacerda et al. (1998) utilized a boundary integral formulation for three-dimensional analyses of thin acoustic barriers over an impedance plane.

In this section, the MacDonald solution for a thin, rigid plane is introduced first as an example of exact solutions, and then another rigorous solution based on the Kirchhoff–Helmholtz equation is presented, which is much easier to understand than the MacDonald solution and the other existing rigorous methods. Finally, an empirical formula used in engineering is discussed.

### 1.2.1 The MacDonald Solution

To calculate the sound field due to a monopole source in proximity to a rigid half-plane, the cylindrical polar coordinate system is used (Li and Wong, 2005). As shown in Figure 1.3, a point monopole source with a source strength of $q_s$ is placed in front of the left surface of the thin plane at $(r_s, \theta_s, y_s)$ with $2\pi \geq \theta_s \geq \pi$, while a receiver located at $(r_r, \theta_r, y_r)$ can be anywhere in the space with $2\pi \geq \theta_r \geq 0$. The straight-line distances from the source and its image to the receivers, $R_1$ and $R_2$, are determined by

$$R_1 = \sqrt{r_s^2 + r_r^2 - 2r_s r_r \cos(\theta_s - \theta_r) + (y_s - y_r)^2} \tag{1.9}$$

and

$$R_2 = \sqrt{r_s^2 + r_r^2 - 2r_s r_r \cos(\theta_s + \theta_r) + (y_s - y_r)^2} \tag{1.10}$$

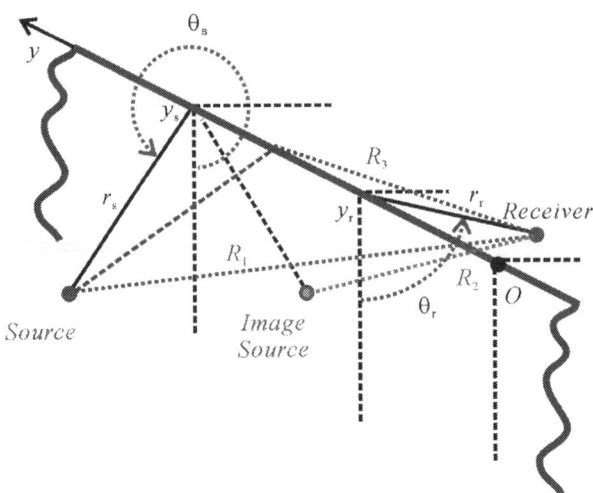

**FIGURE 1.3**
Schematic drawing for calculating the sound diffraction by a semi-infinite plane.

The shortest distance over the barrier from the source to the receiver is called the source-edge-receiver path and can be calculated using

$$R_3 = \sqrt{(r_s + r_r)^2 + (y_s - y_r)^2} \tag{1.11}$$

The total sound field at location $\mathbf{r}$ can be expressed as (Bowman, Senior, and Uslenghi, 1969)

$$p_p(\mathbf{r}) = \frac{k^2 \rho_0 c_0 q_s}{4\pi} \int_{\zeta_1}^{\infty} \frac{H_1^{(1)}(kR_1 + s^2)}{\sqrt{2kR_1 + s^2}} ds + \frac{k^2 \rho_0 c_0 q_s}{4\pi} \int_{\zeta_2}^{\infty} \frac{H_1^{(1)}(kR_2 + s^2)}{\sqrt{2kR_2 + s^2}} ds \tag{1.12}$$

where $k$ is the wave number of the incident wave, $\rho_0$ and $c_0$ are the air density and the sound speed in air, and $H_1^{(1)}(\cdot)$ is the Hankel function of the first kind. $\zeta_1$ and $\zeta_2$ are the limits of the integrals, which are determined by,

$$\zeta_1 = \text{sgn}(\theta_s - \pi - \theta_r)\sqrt{k(R_3 - R_1)} \tag{1.13}$$

and

$$\zeta_2 = \text{sgn}(3\pi - \theta_s - \theta_r)\sqrt{k(R_3 - R_2)} \tag{1.14}$$

where sgn($\cdot$) is the sign function, which takes the value of +1 or −1 depending on the sign of the argument.

At an extremely high frequency, the integral limits can be approximated by infinity. In region I, shown in Figure 1.2, $\theta_s - \pi - \theta_r \leq 0$, $3\pi - \theta_s - \theta_r \leq 0$, so $\zeta_1 \longrightarrow -\infty$, $\zeta_2 \longrightarrow -\infty$, and the total sound field at location $\mathbf{r}$ can be expressed as

$$p_p(\mathbf{r}) \approx \frac{k^2 \rho_0 c_0 q_s}{4\pi} \int_{-\infty}^{\infty} \frac{H_1^{(1)}(kR_1 + s^2)}{\sqrt{2kR_1 + s^2}} ds + \frac{k^2 \rho_0 c_0 q_s}{4\pi} \int_{-\infty}^{\infty} \frac{H_1^{(1)}(kR_2 + s^2)}{\sqrt{2kR_2 + s^2}} ds$$

$$= \frac{j\omega \rho_0 q_s}{4\pi R_1} e^{-jkR_1} + \frac{j\omega \rho_0 q_s}{4\pi R_2} e^{-jkR_2} \quad (1.15)$$

Therefore, the total sound field, $p_p$ in region I, is only composed of the directed sound by the primary source and the reflected sound by the barrier at this high frequency under this circumstance. In region II, $\theta_s - \pi - \theta_r \le 0$, $3\pi - \theta_s - \theta_r \ge 0$, so $\zeta_1 \longrightarrow -\infty$, $\zeta_2 \longrightarrow \infty$, the total sound field at location $\mathbf{r}$ can be expressed as

$$p_p(\mathbf{r}) \approx \frac{k^2 \rho_0 c_0 q_s}{4\pi} \int_{-\infty}^{\infty} \frac{H_1^{(1)}(kR_1 + s^2)}{\sqrt{2kR_1 + s^2}} ds = \frac{j\omega \rho_0 q_s}{4\pi R_1} e^{-jkR_1} \quad (1.16)$$

The total sound field $p_p$ in region II is composed of the directed sound by the primary source at this high frequency. In region III, because $\theta_s - \pi - \theta_r \ge 0$, $3\pi - \theta_s - \theta_r \ge 0$, so $\zeta_1 \longrightarrow \infty$, $\zeta_2 \longrightarrow \infty$, and the total sound field at location $\mathbf{r}$ is zero. This is the dark region and the diffracted sound is almost zero because the sound wavelength at extremely high frequency is so small and the sound waves behave like light rays.

At a very low frequency, $\zeta_1 \longrightarrow 0$, $\zeta_2 \longrightarrow 0$, the total sound pressure at a far-field receiver point ($R_1 \approx R_2$) is approximately

$$p_p(\mathbf{r}) \approx \frac{k^2 \rho_0 c_0 q_s}{4\pi} \int_0^{\infty} \frac{H_1^{(1)}(kR_1 + s^2)}{\sqrt{2kR_1 + s^2}} ds + \frac{k^2 \rho_0 c_0 q_s}{4\pi} \int_0^{\infty} \frac{H_1^{(1)}(kR_2 + s^2)}{\sqrt{2kR_2 + s^2}} ds$$

$$\approx \frac{j\omega \rho_0 q_s}{4\pi R_1} e^{-jkR_1} \quad (1.17)$$

Because of the extremely large wavelength at the low frequency, the sound field behaves like that from a point monopole source without the thin half-plane.

### 1.2.2 The Zhao Solution

Numerical schemes can be used to calculate the sound field with an acoustic barrier. Numerical methods have been used to study the influence of shape and absorbing surface of sound barriers with two-dimensional and three-dimensional boundary element methods (BEMs) (Morgan, Hothersall, and Chandler-Wilde, 1998). Although the BEM can achieve high accuracy, they generally consume large amounts of computation memory and time. Here, an integral equation method is introduced for calculating the sound field diffracted by a rigid barrier on an impedance ground (Zhao, Qiu, and

Cheng, 2015). In this method, a virtual boundary is assumed to model the sound paths above the rigid barrier, and the whole space is divided into two subspaces by this virtual boundary and the rigid barrier.

Figure 1.4(a) shows the rigid barrier on the ground to be investigated and the coordinate system for solving the problem. The rigid barrier is assumed to be infinitely thin and located at $x=0$, $0<y<h$, where $h$ is the height of the rigid barrier, and the ground is represented by $y=0$. The ground can be rigid, soft, or with any impedance. Assume the admittance of the ground is $\beta$. $\beta=0$ denotes the hard ground while $\beta=\infty$ denotes the soft ground. A point monopole source S is located at $\mathbf{r}_s=(x_s, y_s, z_s)$ on the left side of the barrier. The barrier is rigid and infinitely long so the only path for the sound propagation to other side of the rigid barrier is the space above the barrier. A virtual boundary, $\Sigma$, as shown in Figure 1.4(b), is used to model the sound path above the barrier, and the sound field behind the barrier is determined by the boundary conditions on the virtual boundary.

To derive the boundary conditions on the virtual boundary, the whole space is divided into two subspaces as shown in Figure 1.4(b). Space I is the space of $x<0$, $y>0$ with ground $C_1$, barrier $B_1$, virtual boundary $\Sigma$, and quadrant $D_1$, where point S indicates the sound source. Space II is the space of $x>0$, $y>0$ with ground $C_2$, barrier $B_2$, virtual boundary $\Sigma$, and quadrant $D_2$. $C_1$ and $C_2$ denote the ground on the left and right sides of the barrier, respectively, $B_1$ and $B_2$ are the left and right side of the rigid barrier, and the diameters of $D_1$ and $D_2$ are infinite.

Following the Kirchhoff–Helmholtz equation (Nelson and Elliott, 1992), the sound field at location $\mathbf{r}$ in space I can be described as,

$$p^{(\mathrm{I})}(\mathbf{r}) = p_i(\mathbf{r}) + \iint_{B_1+\Sigma+D_1+C_1} \left[ G(\mathbf{r}\,|\,\mathbf{s}) \frac{\partial p^{(\mathrm{I})}(\mathbf{s})}{\partial \mathbf{n}} - p^{(\mathrm{I})}(\mathbf{s}) \frac{\partial G(\mathbf{r}\,|\,\mathbf{s})}{\partial \mathbf{n}} \right] dS \qquad (1.18)$$

where $\mathbf{n}$ is the unit vector normal to the boundaries shown in Figure 1.4(b), $\mathbf{s}$ is the position vector on the boundaries and $p_i(\mathbf{r})$ is the sound pressure at $\mathbf{r}$

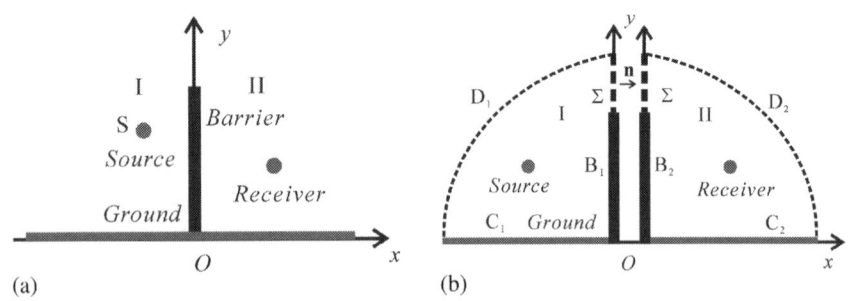

(a)                                              (b)

**FIGURE 1.4**
(a) Two subspaces of a sound barrier on the ground and the coordinate system, (b) the boundaries (including the virtual boundaries) used in the sound field calculation.

generated by the point monopole source in free field, which can be obtained with $p_i(\mathbf{r}) = j\omega\rho_0 q_s G(\mathbf{r}|\mathbf{r}_s)$. $j$ is the imaginary unit, $\omega$ is the angular frequency, $\rho_0$ is the air density, $q_s$ is the source strength of the point monopole source, and $G(\mathbf{r}|\mathbf{r}_s)$ is the Green function.

The sound field must satisfy the rigid boundary conditions of barrier $B_1$ and the impedance boundary conditions of the ground $C_1$ simultaneously,

$$\left.\frac{\partial p(\mathbf{s})}{\partial \mathbf{n}}\right|_{B_1} = 0 \tag{1.19}$$

$$\left.\frac{\partial p(\mathbf{s})}{\partial \mathbf{n}}\right|_{C_1} + jk\beta(\omega)p(\mathbf{s})\big|_{C_1} = 0 \tag{1.20}$$

For convenience of calculation, the Green function can be selected so that

$$\left.\frac{\partial G(\mathbf{r}|\mathbf{s})}{\partial \mathbf{n}}\right|_{B_1} = \left.\frac{\partial G(\mathbf{r}|\mathbf{s})}{\partial \mathbf{n}}\right|_{\Sigma} = 0 \tag{1.21}$$

$$\left.\frac{\partial G(\mathbf{r}|\mathbf{s})}{\partial \mathbf{n}}\right|_{C_1} + jk\beta(\omega)G(\mathbf{r}|\mathbf{s})\big|_{C_1} = 0 \tag{1.22}$$

By combining the boundary conditions and the fact that the integration over $D_1$ tends to be zero with the infinitely large diameter, Equation (1.18) becomes

$$p^{(\mathrm{I})}(\mathbf{r}) = j\omega\rho_0 q_s G(\mathbf{r}|\mathbf{r}_s) + \int_{-\infty}^{\infty}\int_{h}^{\infty} G(\mathbf{r}|0,y,z)\left.\frac{\partial p^{(\mathrm{I})}(x,y,z)}{\partial x}\right|_{x=0} \mathrm{d}y\mathrm{d}z \tag{1.23}$$

Similarly, the sound field at location $\mathbf{r}$ in space II can be described as,

$$p^{(\mathrm{II})}(\mathbf{r}) = -\int_{-\infty}^{\infty}\int_{h}^{\infty} G(\mathbf{r}|0,y,z)\left.\frac{\partial p^{(\mathrm{II})}(x,y,z)}{\partial x}\right|_{x=0} \mathrm{d}y\mathrm{d}z \tag{1.24}$$

By using the continuation conditions on the virtual boundary $\Sigma$,

$$p^{(\mathrm{I})}(\mathbf{r})\big|_{x=0,y>h} = p^{(\mathrm{II})}(\mathbf{r})\big|_{x=0,y>h} \tag{1.25}$$

$$\left.\frac{\partial p^{(\mathrm{I})}(\mathbf{r})}{\partial x}\right|_{x=0,y>h} = \left.\frac{\partial p^{(\mathrm{II})}(\mathbf{r})}{\partial x}\right|_{x=0,y>h} \tag{1.26}$$

it has

$$-2\int_{-\infty}^{\infty}\int_{h}^{\infty} G(0,y,z|0,y,z)f(y,z)\mathrm{d}y\mathrm{d}z = j\omega\rho_0 q_s G(0,y,z|\mathbf{r}_s) \tag{1.27}$$

where

$$f(y,z) = \left.\frac{\partial p^{(I)}(x,y,z)}{\partial x}\right|_{x=0,y>h} = \left.\frac{\partial p^{(II)}(x,y,z)}{\partial x}\right|_{x=0,y>h} \tag{1.28}$$

is the pressure gradient distribution on the virtual boundary. The sound field at any location **r** in the space can be calculated with Equations (1.23) and (1.24) after $f(y, z)$ is obtained by solving Equation (1.27).

The Green function that satisfies Equation (1.21) is the sound field generated by a point monopole source in front of an infinitely large rigid plane at $x=0$, which can be obtained easily with the image source method. The Green function required in Equation (1.22) is the sound field generated by a point monopole source above a finite impedance ground at $y=0$. This problem has been widely studied in the past few decades and various theoretical and experimental results have been published (Rudnick, 1947; Ingard, 1951). For example, the Weyl–Van der Pol formula can be utilized in the calculation (Wong and Li, 2001), and the Green function that satisfies Equations (1.22) and (1.23) can be written as

$$G(\mathbf{r}\,|\,\mathbf{s}) = \frac{e^{-jk|\mathbf{r}-\mathbf{s}|}}{4\pi\,|\,\mathbf{r}-\mathbf{s}\,|} + \frac{e^{-jk|\mathbf{r}-\mathbf{s}_1|}}{4\pi\,|\,\mathbf{r}-\mathbf{s}_1\,|} + Q\frac{e^{-jk|\mathbf{r}-\mathbf{s}_2|}}{4\pi\,|\,\mathbf{r}-\mathbf{s}_2\,|} + Q\frac{e^{-jk|\mathbf{r}-\mathbf{s}_3|}}{4\pi\,|\,\mathbf{r}-\mathbf{s}_3\,|} \tag{1.29}$$

where $\mathbf{s}_1=(-x, y, z)$, $\mathbf{s}_2=(x,-y, z)$ and $\mathbf{s}_3=(-x,-y, z)$, being the image source locations of $\mathbf{s}=(x, y, z)$ regarding the barrier plane, the ground, and the origin, respectively. $Q$ is the spherical wave reflection coefficient of the ground. $Q=1$ when the ground is hard ($\beta=0$), while $Q=-1$ when the ground is soft ($\beta=\infty$). For a finite impedance ground, $Q=R_p+(1-R_p)F$, where $R_p$ is the plane wave reflection coefficient and $F$ is an integral dependent on many factors such as the frequency, the distance between the image source and the receiver, the admittance of the ground, and the reflection angle (Rudnick, 1947).

From the above derivation process, the sound pressure gradient distribution on the virtual boundary can be determined by constructing an appropriate Green function and enforcing the continuity conditions, and then the sound field behind the barrier can be derived rigorously. This method is not only valid in the full frequency range and the whole space in principle, but it is also easy to be understood for most acoustic researchers and engineers because it starts from the basic Kirchhoff–Helmholtz integral equation in acoustics, and the derivation process is straightforward and clear. The derivation shows that the whole space above the rigid barrier is important for the calculation instead of just considering the diffracted wave from the edge of the barrier.

Although this method is a rigorous solution and holds in the whole space, it is hard to solve Equation (1.27) analytically, so the discretization operation in the derivation of the boundary condition on the virtual boundary is necessary. The integration range is large when the source and receiver are far away from the barrier, so faster computers with large memories are needed for calculation of the whole sound field (Zhao, Qiu, and Cheng, 2015).

### 1.2.3 The Kurze and Anderson Formula

The acoustic performance of a sound barrier can be determined based on the formulae obtained by fitting the measurement data. Maekawa (1968) published a comprehensive set of measured attenuation data of a thin, rigid barrier for different source and receiver locations. In his measurements, a pulsed tone of short duration was used as the noise source, and the experimental data was presented in a chart that plotted the attenuation against a single parameter known, as the Fresnel number. The Fresnel number is the numerical ratio of the path difference in distance between the diffracted path and the direct path of sound to the half of a sound wavelength. Kurze and Anderson (1971) proposed an improved empirical formula for the sound barrier attenuation based on the diffraction theory from Keller (1962) and the existing experimental data from Maekawa.

As shown in Figure 1.3, for a point monopole source located at $(r_s, \theta_s, y_s)$ in front of the left surface of a thin plane and a receiver located at $(r_r, \theta_r, y_r)$, the straight-line distance from the source to the receiver can be obtained by

$$R_1 = \sqrt{r_s^2 + r_r^2 - 2r_s r_r \cos(\theta_s - \theta_r) + (y_s - y_r)^2} \tag{1.30}$$

The shortest distance over the barrier from the source to the receiver (the source-edge-receiver path) can be calculated with

$$R_3 = \sqrt{(r_s + r_r)^2 + (y_s - y_r)^2} \tag{1.31}$$

Let $\lambda$ be the wavelength of the sound at the frequency of interest, the corresponding Fresnel number of the source and the receiver with the barrier is

$$N = \frac{R_3 - R_1}{\lambda / 2} \tag{1.32}$$

The insertion loss brought by a rigid, thin, infinitely long barrier to the source and receiver can be simply estimated with (Bies, Hansen, and Howard, 2018):

$$IL = 10 \log_{10}(3 + 20N) \tag{1.33}$$

or, more accurately, with the Kurze and Anderson formula as

$$IL = 5 + 20 \log_{10} \frac{\sqrt{2\pi N}}{\tanh \sqrt{2\pi N}} \tag{1.34}$$

Equation (1.32) shows that the Fresnel number is proportional to the difference in distance between the diffracted path and the direct path of sound, so for a source or receiver close to the barrier or the receiver close to the boundary of the bright and shadow zones, the Fresnel number is small, resulting in a small insertion loss. Because the Fresnel number becomes smaller at a lower frequency, the insertion loss of a barrier is low in the low-frequency range. Figure 1.5 shows the insertion loss calculated by using Equations (1.33) (solid line) and (1.34) (dashed line) for a 4 m high, thin, rigid barrier with a point monopole source 2 m away from the barrier at one side and a receiver 3 m away at other side. Both the source and receiver are at the same height of 1 m, and their direct distance is 5 m. It is clear that the difference between the two formulae is small.

Equation (1.33) is only valid for $N > 0$. The difference between the calculated value and the Maekawa's data becomes larger for $N < 0.5$, and the maximum bias of about 1.5 dB appears at $N = 0.1$. In 2001, an improved formula was proposed to provide greater accuracy for locations of the source or receiver close to the barrier or for the receiver close to the boundary of the bright and shadow zones by taking into account the Fresnel number of the image source to the rigid barrier and the receiver (Menounou, 2001). Different types of waves from different sources, such as plane waves, cylindrical waves (coherent line sources), and spherical waves (point monopole sources), were treated separately in the improved formula. In practical applications, the reflections from the ground, the diffraction around the vertical ends of finite length barriers, and the thickness of the barrier all need to be taken into consideration, so the formulae are getting more complicated. More details can be found in the literature (Li and Wong, 2005; Attenborough, Li, and Horoshenkov, 2007; Bies, Hansen, and Howard, 2018).

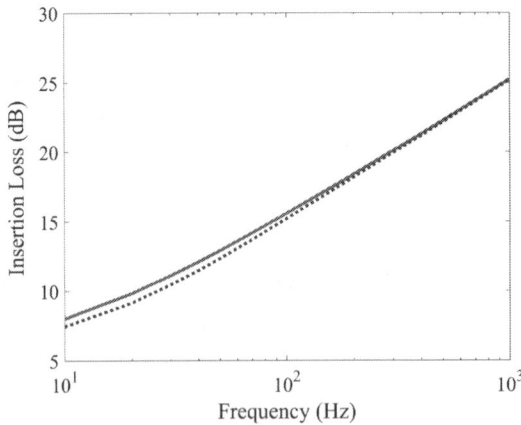

**FIGURE 1.5**
The insertion loss calculated by using Equations (1.33) (solid line) and (1.34) (dashed line) for a 4 m high barrier with a point monopole source 2 m away from the barrier at one side and a receiver 3 m away at other side. Both the source and receiver are at the same height of 1 m, and their direct distance is 5 m.

## 1.3 Active Sound Barriers

As discussed in Section 1.2 and shown in Figure 1.5, the insertion loss of passive barriers is usually poor in the low-frequency range due to the large wavelength of the sound. To overcome this problem, active noise control (ANC) has been applied on passive barriers by installing a number of secondary sources (loudspeakers) near the edges of the barriers to constitute hybrid passive and ANC systems, called active sound (noise) barriers (ANB). The ANC system acts to cancel the diffraction sound around the barrier edge, which is believed to contribute significantly to the diffracted sound field in the dark area behind the barrier. If the diffracted sound around a barrier edge is canceled, the diffracted sound field in the dark area behind the barrier should be attenuated. The effect of the ANC systems is equivalent to increasing the height of the passive sound barriers.

The first three-dimensional multiple-channel ANB systems were developed by Omoto and Fujiwara (1993), who applied a multichannel ANC system to cancel the noise around a semi-infinite barrier. Based on the numerical simulations and experiments for the specific configuration studied, it is found that, when the separation distance between the error sensors is less than half the wavelength, the active control system works effectively and the attenuation is greater when the secondary sources are nearer to the primary source.

Four years later in 1997, they further investigated the influence of ground reflection on the active control system through outdoor tests, and the experimental results show that the active system works normally with about 6 dB excess attenuation over the passive barrier's insertion loss at the receiver 50 m away in the 125 Hz one-octave band, which is equivalent to increasing the height of the barrier from 2 m to 5 m (Omoto et al., 1997). The numerical simulation and experimental results from Guo and Pan (1998) show that, although the ground reflection decreases the noise reduction performance of the ANB system, the system is still effective to decrease the diffracted sound all the time.

The influence of the geometrical shape of the secondary sources was investigated, and the arc-type arrangement was found to be more effective than the linear-type one with the same number of the secondary sources (Shao, Sha, and Zhang, 1997). Based on this discovery, Yang and Gan (2001) discussed the methods to choose the cost function for the ANB systems with the secondary sources being arranged as the arc-type. The effects of secondary source distribution and density were studied, and a double row arrangement was shown to be more effective in attenuating noise in the shadow zone rather than a single row arrangement (Liu and Niu, 2008). Chen et al. (2011) proposed to use specially designed unidirectional secondary sources to further increase the performance of ANB systems, while Hart and Lau

(2012) investigated ANC with linear secondary source and sensor arrays for the noise barrier.

In active sound barrier systems, the locations of the error microphones are usually on the top of the barriers, and the locations of the secondary sources are between the primary source and error microphones. This arrangement makes the installation of the active systems difficult in practice. Niu et al. (2007) examined the optimal positions of the error sensors in an ANB system, and found that it would be better to locate the error sensors above the secondary sources, with an optimal distance between the error sensors and the secondary sources of 8 cm for the system with the secondary sources on the top of the noise barrier and the error microphones near the secondary sources. In the same year, Han and Qiu (2007) proposed to use the sound intensity as the cost function for active control to enhance the insertion loss of an ANB system in the far field. Later, Lau and Tang (2009) investigated system configuration and pressure gradient control for active sound barriers and found that the method of controlling the pressure gradient is more efficient than that of controlling the sound pressure.

The virtual sensor approach was also applied to an ANB system, which used near-field error microphones to obtain the far-field error signals so as to control the far-field noise (Berkhoff, 2005). The simulations show that the performance using the virtual far-field error signals is similar to that using the far-field error signals. Recently, Qiu and Zhao (2015) proposed controlling the directivity of the sound diffraction through the barrier edge so that the sound field is reduced in certain directions. Numerical simulations show that their proposed directivity control method can provide better performance than that of the local control in terms of the far-field sound reduction.

The first prototype practical ANB system in engineering was demonstrated in 2004, when a 20 m long product-type active soft-edge noise barrier was implemented along a highway and 2 dB excess insertion loss was gained by the active sound barrier system on-site (Ohnishi et al., 2004). Active sound barriers can not only be applied along traffic roads for traffic noise control but can also be applied to other noise sources. For example, Figure 1.6 shows two prototype ANB systems, one is for traffic noise control where an array of loudspeakers is located on the top of a transparent passive barrier to reduce the sound diffraction over the top edge of the barrier, while the other one is for transformer noise control where a 15-channel ANC system is used vertically with a high wall around the transformer to reduce the sound diffraction along the edge of the side opening (Zou et al., 2014). On-site measurement data shows that an average noise reduction of about 0.3–4.3 dBA below 400 Hz is obtained on the transformer plant boundary with the system shown in Figure 1.6(b).

In this section, the principle and methods about active sound barriers are introduced first, and then the secondary sources and error-sensing strategies for active sound barriers are presented. Finally, the implementation issues, the challenges, and future directions are discussed.

**FIGURE 1.6**
(a) A prototype active sound barrier system for traffic noise control that uses an array of loud-speakers on the top of a transparent passive barrier to reduce the sound diffraction over the top edge of the barrier, (b) a prototype active sound barrier system for transformer noise control where a 15-channel ANC system is used vertically together with the high wall around the transformer to reduce the sound diffraction along the edge of the side opening.

## 1.3.1 Principle

The objective of active sound barriers is to control the sound diffraction along the edge of the barriers by installing many secondary sound sources to interact with the "virtual sources" of the diffracted field. If the sound pressure at the vicinity of the edge has a dominant effect on the diffracted field, then continuous cancellation of the sound pressure over an arbitrary region near the diffraction edge is required to control sound diffraction completely. In practice, continuous cancellation is difficult to implement, so discretization has been applied to carry out cancellation at multiple points along the diffraction edge. There are several questions to be answered before implementation. For example, where to put these secondary sources and error sensors, and how to set up their strength?

Figure 1.7 shows a simple feedforward active sound barrier system that consists of a number of reference sensors, secondary sources, and error sensors. The reference sensors are usually located before the secondary sources so that the primary sound field encounters the reference sensors first in its propagation, while the error sensors are usually behind the secondary sources to pick up the sound field after the primary sound wave passes the secondary sources. The architecture of active sound barrier systems can be feedforward or feedback. For the feedforward control systems, the reference sensors pick up the information from the primary noise source first, and then the controller processes the information to drive the secondary sources to reduce the diffraction along the edge. For the feedback control systems, the information obtained at the error sensors is used to drive the secondary sources to change the response of the secondary sources to the primary sources. Both architectures have been successfully used in active sound barrier applications. To obtain better noise reduction performance, both reference sensors and error sensors are used to construct an adaptive

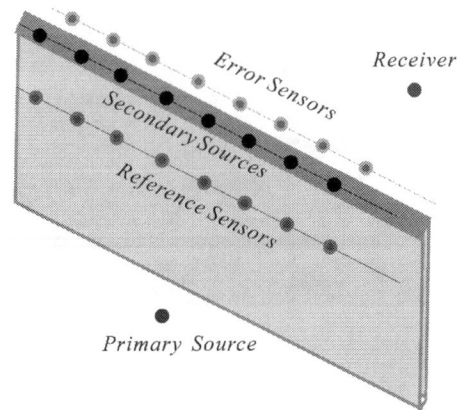

**FIGURE 1.7**
A simple active sound barrier system that installs a number of secondary sources on the edge of the barrier evenly to reduce the diffraction from the passive barrier to a primary noise source; the active sound barrier system can be a feedforward control system that uses the reference sensors, a feedback control system that uses the error sensors, or an adaptive feedforward system that uses both the reference and error sensors.

feedforward control system, which processes the information from the reference sensors and uses the information from the error sensors to adjust the control coefficients of the controller.

The sound pressure at location $\mathbf{r}$ generated by the primary noise source located at $\mathbf{r}_p$ can be written as

$$p_p(\mathbf{r}) = Z_p q_p \tag{1.35}$$

where $q_p$ is the primary source strength, and $Z_p$ is the transfer impedance from the source location $\mathbf{r}_p$ to the field location $\mathbf{r}$, which can be obtained with the MacDonald solution (Section 1.2.1), the Zhao solution (Section 1.2.2), other numerical solutions and commercial software, or from measurements.

Similarly, the sound field at location $\mathbf{r}$ from the secondary sources can be obtained as,

$$p_s(\mathbf{r}) = \sum_{l=1}^{L} Z_{sl} q_l = \mathbf{Z}_s \mathbf{q}_s \tag{1.36}$$

where $L$ is the number of the secondary sources, $q_l$ is the strength of the $l$th secondary source, $Z_{sl}$ is the transfer impedance from the $l$th secondary source located at $\mathbf{r}_l^s$ to the field location $\mathbf{r}$, $\mathbf{Z}_s = [Z_{s1}, Z_{s2}, ..., Z_{sL}]$, and $\mathbf{q}_s = [q_1, q_2, ..., q_L]^T$. $Z_{sl}$ can be calculated in the same way as that for $Z_p$.

The total sound field from both the primary source and secondary sources can be expressed as

$$p_t(\mathbf{r}) = p_p(\mathbf{r}) + p_s(\mathbf{r}) \tag{1.37}$$

The cost function can be defined according to the objective of the control. Take the sum of the squared sound pressure at a number of error sensor locations as an example,

$$J_p = \sum_{m=1}^{M} |p_t(\mathbf{r}_{e,m})|^2 + \beta \mathbf{q}_s^H \mathbf{q}_s \tag{1.38}$$

where $M$ is the number of the error sensors, $\mathbf{r}_{e,m}$ is the location of the $m$th error sensor, $\beta$ is a positive real number, which can be set to an appropriate value to constrain the secondary source output power, and the superscript H denotes the conjugate transpose operation. The optimal secondary source strength can be obtained by minimizing the above cost function and can be expressed in matrix form as

$$\mathbf{q}_{so} = -(\mathbf{Z}_s^H \mathbf{q}_s + \beta \mathbf{I})^{-1} \mathbf{Z}_s^H \mathbf{P}_p \tag{1.39}$$

where $\mathbf{P}_p = [p_p(\mathbf{r}_{e,1}), p_p(\mathbf{r}_{e,2}), \ldots, p_p(\mathbf{r}_{e,M})]^T$ is the sound pressure vector generated by the primary source.

After obtaining the optimal secondary source strength, it can be substituted back to Equation (1.37) to calculate the total sound pressure amplitude with active control. The excess insertion loss due to the active control system can be expressed as

$$\mathrm{IL} = 20 \log_{10} \left| \frac{p_p(\mathbf{r})}{p_t(\mathbf{r})} \right| \tag{1.40}$$

where $p_p(\mathbf{r})$ and $p_t(\mathbf{r})$ are the sound pressure without and with active control at location $\mathbf{r}$.

Equations (1.35–1.39) can be used to calculate the insertion loss of an ANB system in the frequency domain. It can also be used to investigate the configuration of the secondary sources as well as the error-sensing strategies (Sections 1.3.2 and 1.3.3). These equations can provide an upper limit of the noise reduction performance of a designed ANB system, whose performance is also affected by the quality of the reference signals, the control hardware and algorithms, and the electroacoustic properties of the secondary sources (Section 1.3.4).

## 1.3.2 Secondary Sources for Active Sound Barriers

The active sound barrier first proposed was based on the concept of the cancellation of sound pressure at a diffraction edge, which behaved like a virtual source for the diffracted field (Omoto and Fujiwara, 1993). As presented by Omoto and Fujiwara, the concept of a "virtual source" is only an interpretation of the approximate expression of the far field from the edge. Because the sound pressure at the vicinity of the edge has a dominant effect on the diffracted field, some level of control of sound propagation could be gained if there is a continuous cancellation of the sound pressure over an arbitrary region near the diffraction edge. The existing simulation and experiment results have been reported to support the thought (Omoto et al., 1997; Ohnishi et al., 2004; Liu and Niu, 2008; Chen et al., 2011; Zou et al., 2014).

Based on the numerical simulations and experiments for a specific configuration, Omoto and Fujiwara (1993) investigated the arrangement of the secondary sources, and they found that the attenuation increases when the secondary sources are nearer to the primary source, wherein the wave front of the canceling wave resembles that of the primary source near the diffraction edge. A few years later, the influence of the geometrical shape of the secondary sources was investigated, and the numerical simulations show that the arc-type arrangement is more effective than the linear-type one with the same number of the secondary sources (Shao, Sha, and Zhang, 1997). Unfortunately, these statements are only applicable to the barriers that are close to the primary noise sources. As pointed out later by Yang and Gan (2001), the main mechanism for active sound barriers is related to the destructive interference near the diffraction edge caused by wave-front matching, and the arc-type arrangement of secondary sources is not generally a better choice than the other arrangement.

To have perfect sound cancellation, the control sound field generated by the secondary sources should match the diffraction field caused by the barrier in the "dark" region. According to the Huygens' principle, every point at a wave front should be considered as the source of the secondary wavelets with a speed equal to the speed of the waves, so a complete cancellation of the diffraction sound field by the barrier to the noise source requires an infinite number of secondary sources distributed over the entire transmission path above the barrier with their strengths opposite to the strengths of secondary wavelets of the diffraction sound wave front. As shown in Section 1.2, the distribution of the diffraction sound field depends on the locations and properties of the primary source and the barrier. For example, if the primary source is a line source instead of a point monopole source, the diffraction sound field is different. Even with the same point primary source, the distance between the source and barrier affects the diffracted sound field. Therefore, the diffraction sound field to be controlled by the secondary sources varies in different situations, and an optimal secondary source configuration for perfect cancellation under all situations might not exist.

Although the diffraction sound fields are different for primary noise sources at different locations, a high passive sound barrier usually can have larger insertion loss than a low one, so it might be feasible if the secondary sound sources can be designed to generate a secondary sound field that interacts with the primary sound field just like that from a high passive sound barrier. Even though this results in imperfect cancellation, increasing the equivalent height of a passive sound barrier in the low-frequency range is a main demand, arising from practice, for the active sound barriers. Unfortunately, even this partial cancellation has many challenges. Some research work has been carried out. For example, Liu and Niu (2008) showed that arranging the secondary sources in a double row on the edge of the barrier is more effective in attenuating noise in the shadow zone than that from a single row arrangement, Chen et al. (2011) designed a special kind of unidirectional secondary source to increase the performance of ANB systems, and Hart and Lau (2012) proposed to use a linear array of secondary sources and a perpendicular linear array of error sensors above the top of a sound barrier to reduce the diffraction wave.

Figure 1.8(a) shows the active sound barrier with unidirectional secondary sources proposed by Chen et al. (2011), where the unidirectional secondary source is constituted by two monopoles at high and low positions with a distance of $d$. The complex strengths of the two elements, $A_1$ and $A_2$ are defined by (Boone and Ouweltjes, 1997),

$$A_1 = \frac{1}{2}Q\left(\alpha + \frac{2}{jkd}\right), \quad A_2 = \frac{1}{2}Q\left(\alpha - \frac{2}{jkd}\right) \tag{1.41}$$

where $Q$ is the common source strength factor of the two elements, and $\alpha$ is a parameter for phase adjustment. The directivity of the source approaches that of a tripole when $\alpha$ equals to 1.0, and its directivity pattern is between tripole and monopole when $\alpha$ varies in the range of $[1, \infty)$. In their experiments, two loudspeakers with similar characteristics were used to constitute a unidirectional source. The distance between the centers of the two elements is about 0.15 m. The value of $\alpha$ is adjusted to 2.0 by using a phase shift circuit. The calculated and measured directivities of the unidirectional source on a circle with a radius of 1.0 m at 300 Hz are shown in Figure 1.8(b), where the

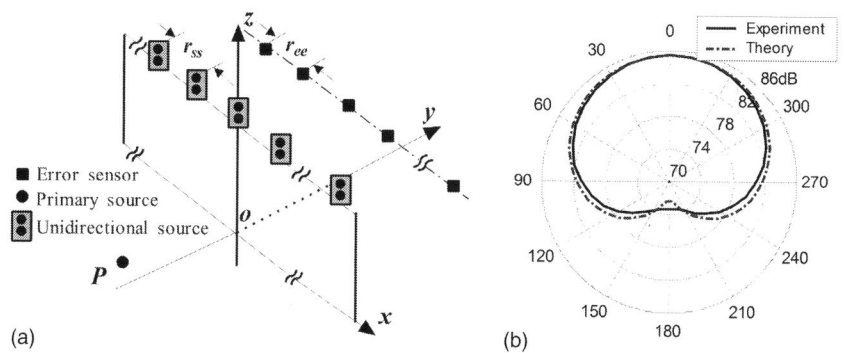

**FIGURE 1.8**
(a) An active noise barrier with unidirectional compound secondary sources, (b) the calculated and measured directivities of the unidirectional compound source on a circle with a radius of 1.0 m at 300 Hz.

unidirectional source exhibits a tripole-like pattern. The experiment results show that the extra insertion loss difference between that using the unidirectional secondary source control system and the secondary monopole source control system is more than 2 dB in the dark region due to the better match of the secondary sound field generated by the unidirectional sources in the dark region.

The effects of the position of the secondary sources on the performance of the active noise barriers have been investigated with two-dimensional numerical simulations (Tao, Zou, and Qiu, 2010). Figure 1.9(a) shows the schematics of a two-dimensional noise barrier, where the rigid noise barrier with a height of 1.8 m is located at $x=0$ and the ground at $y=0$ is perfectly reflective. The primary line source vertical to the $x$–$y$ plane is located at $(-6.0$ m, 0.1 m) and $L_1$ $(x=6.0$ m, 0 m $<y<3.5$ m) is an observation line in the "dark region". Four positions with equal horizontal distances to the barrier, $S_1$ $(-0.5,$ 1.8), $S_2$ $(-0.5$ 0.1), $S_3$ $(0.5, 1.8)$, and $S_4$ $(0.5, 0.1)$ are chosen for the investigation of the secondary source placement. The total sound potential energy in the region of $(6$ m $<x<16$ m, 0 m $<y<1.8$ m) is used to evaluate the performance of the active sound barrier. The results in Figure 1.9(b) show that placing the secondary source along the edge gives the best noise reduction performance, and placing the secondary source near the ground and on the same side of the primary source can also provide acceptable performance.

### 1.3.3 Sensing Strategies for Active Sound Barriers

Assume that sufficient number of the secondary sources can be used with a passive sound barrier to constitute an active sound barrier to increase the insertion loss of the barrier. The question is how to set up the strength of these secondary sources. Ideally, it is desired that the active sound barrier can behave like an infinitely high, rigid, passive barrier, which means that there

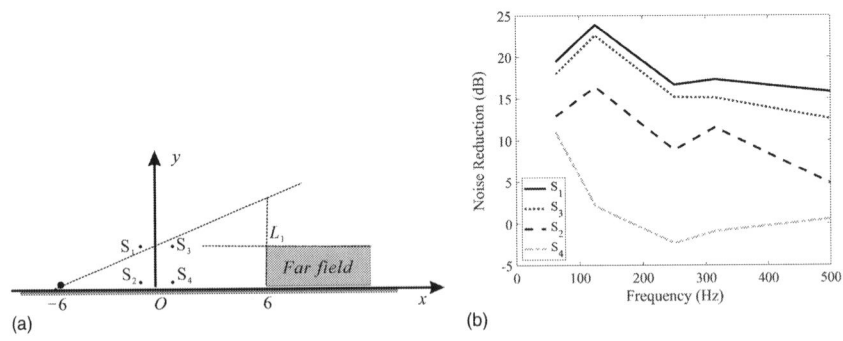

**FIGURE 1.9**
(a) Locations of the secondary sources in the two-dimensional active sound barrier, (b) the performance of the active sound barrier with the secondary sources at different positions (from top to bottom, S1 – solid line, S3 – dotted line, S2 – dashed line, S4 – dash-dotted line).

is no sound energy propagating above the edge of the barrier. Therefore, a natural approach is to have an infinite number of sound intensity error sensors installed evenly in the space above the edge of the passive barrier to monitor the sound energy passing over the barrier, and the strength of the secondary sources is tuned to make the mean active sound intensity passing over the barrier edge zero. Unfortunately, this method is not practical because it is hard to install so many sensors on the top above the edge of the barrier.

In the early research on active sound barriers, it is believed that the sound pressure at the vicinity of the edge has a dominant effect on the diffracted field, so Omoto and Fujiwara (1993) installed the error sensors on the edge of the barrier, and their numerical and experimental results show that the separation of distance between the error sensors should be less than a half the wavelength of the sound to be controlled in order to have effective control. In their research, the error microphones were placed on top of the barriers and the secondary sources were between the primary source and error microphones. In practice, this makes installation of the active system difficult. To make it more practical, Niu et al. (2007) examined the optimal positions of the error sensors in an active sound barrier system for the system with the secondary sources on top of the sound barrier and the error microphones near the secondary sources. They found that it would be better to locate the error sensors above the secondary sources with an optimal distance between the error sensors and the secondary sources of 8 cm in order to obtain the greatest excess insertion loss in the "dark" region.

To further increase the insertion loss in the far field, Han and Qiu (2007) proposed to use the sound intensity as the cost function for active control. They found that minimizing the near-field sound intensity at discrete locations along the edge of the passive sound barrier, the sound intensity is reduced as is the sound pressure in the far field. The results from their offline experiments for a 3-channel control system show that the active sound intensity control is able to provide better far-field noise reduction than the squared sound pressure control. Later, Lau and Tang (2009) proposed to use a perpendicular linear array of error sensors above the top of a noise barrier to reduce the diffraction wave with a linear array of secondary sources. Their simulation results show that the extra insertion loss increased with the number of the lines of error sensors (i.e., the line of control) and locating the line of error sensors closer to the barrier edge provides better noise reduction performance due to better matching to the diffraction sound field.

In most applications, the objective of active sound barriers is to reduce far-field noise in the "dark" region, but it is impractical to locate the error sensors in the far field for the active sound barrier systems because of the need to manage a compact system and the consideration of the stability and tracking ability of the control system. Berkhoff (2005) applied the virtual sensor method to an active sound barrier system, which used near-field error microphones to obtain the far-field error signals. The simulation results show that the performance of the system using the virtual far-field error signals can gain the same performance as that using the far-field error signals. He found that the contributions of the primary sources and the secondary sources should be treated separately to use the virtual sensors, and a system with all transducers placed near a noise barrier seems to be relatively robust with respect to changes in the propagation characteristics between the sound barrier and the target region.

In some situations, only the diffraction sound in certain directions is required to be reduced. Qiu and Zhao (2015) proposed controlling the directivity of the sound diffraction through the barrier edge so that the sound field is reduced only in certain specified directions. In their simulations, 26 secondary sources distributed evenly along the barrier edge with a space of 0.2 m are used to control the diffraction sound field generated by a primary point monopole source. The cost function used is the sum of the squared pressure at 71 evenly distributed points for each direction from 0 m to 7 m with a space of 0.1 m. The numerical simulation results show that their proposed directivity control method can provide better performance than that of local control in terms of the far-field sound reduction. Figure 1.10 shows the primary sound field and the sound field with the directivity control at both 30° and 60° for the 250 Hz tonal sound.

### 1.3.4 Implementation Issues

Active sound barriers have been studied for nearly three decades. Several prototypes have been installed along traffic roads and tested, but there is still no mature commercial product. There are many challenges for practical implementation. The first is the noise reduction performance. Although passive sound barriers have been widely used in practice for traffic noise control, they are only effective for the "dark" region. Active sound barriers act as if they are increasing the equivalent height of the passive barriers, but they are only effective in the low-frequency range. Usually, the excess insertion loss obtained from the active sound barriers is only 2–3 dB, which makes the solution not very attractive to industry.

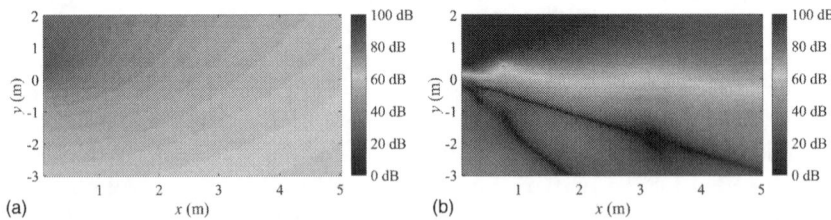

**FIGURE 1.10**
(a) The diffraction sound field generated by a primary point monopole source without control at 250 Hz, (b) the residual sound field at 250 Hz with the directivity control at 30° and 60°.

The second challenge is the cost. Passive sound barriers built with common materials, such as cement, steel or plastic panels, are usually at low cost. Active sound barriers use loudspeakers, microphones, and electronic control circuits or digital signal processors to build complicated control systems, which are often at high cost. Passive sound barriers can be installed outdoors easily and can function for years with little maintenance, while active sound barriers require an electrical power supply all the time, and the lifetime of the electronic and acoustic components might be only a few years in outdoor environments. Therefore, the maintenance cost for active sound barriers is much higher than for passive ones.

To further increase the noise reduction performance of active sound barriers, there are still many problems that need to be solved. First, a comprehensive understanding of the mechanisms of the active sound barrier systems is needed for further optimization of the system parameters, such as the locations and types of the secondary sources and the error sensors. Although the related studies have been improving the system performance continuously, the noise reduction mechanisms of the active sound barriers are still not very clear. The sound field and acoustic power flow from different regions around an active noise barrier have been analyzed based on a two-dimensional model (Chen, Min, and Qiu, 2013), and the numerical results show that the dominant mechanism is the energy reflection from the shadow zone to the adjacent region by the secondary sources when the distance between the primary source and the secondary source is large. But the diffracted sound field and its interaction with the secondary sound field generated by the secondary sources in a three-dimensional space are still not clear.

Although many researchers have investigated the error-sensing strategies for active sound barriers such as minimizing the mean sound intensity, using microphone arrays or virtual sensors, there is still no effective way to construct a compact system to guarantee the reduction of the diffraction sound in the "dark" region. Furthermore, for the feedforward control system to be applied along traffic roads, reference sensors have to be used to pick up the primary noise from many directions. Ideally, each secondary source needs to control the primary noise from many directions, which implies that many reference sensors at different directions should be used. To have better coherence between the reference signal and the error signal for one control channel, it is desirable that the reference sensor can be located as close to the error sensor as possible, but this brings the problem of causality for the controller and also the problem of acoustic feedback from the secondary sources to the reference sensors.

In order to work properly for traffic noise, just like passive sound barriers, active sound barriers can be hundreds of meters long. Assume several secondary sources are used per meter along the top of a passive sound barrier, hundreds or thousands of secondary sources are needed in practice, so it is impossible to apply centralized control (Zhao et al., 2017). At present, prototype active sound barrier systems adopt decentralized feedback or feedforward architectures. This means multiple independent single-channel subsystems are used in the whole system (Ohnishi et al., 2004; Zou, Lu, and Qiu, 2010). Feedback control systems have the advantages of compactness and easy implementation; however, the performance of feedback control is limited for two reasons. One is that the error sensors have to be placed very close to that the secondary sources to maintain the stability of the system, and the other is that the feedback system suffers from the waterbed effect.

The implementation of the decentralized feedforward control systems can amend the problems. The simulation and experiment results demonstrate that the decentralized feedforward control systems with predefined control filter parameters and adaptive control systems both work effectively. The challenges are how to maintain the stability of the multiple independent single-channel systems and how to set up and tune each controller. The performance of the systems that apply predefined control filter parameters is not as good as that use adaptive control, but it is more compact due to the fact that the error sensors are no longer needed in the operating stage. It is also more stable than the latter.

In addition to above issues, the change of meteorology effects such as temperature and wind can also affect the noise reduction performance and the stability of the control systems. To maintain the noise reduction performance under most circumstances, the system needs to be adaptive so that the control parameters can be adjusted according to the environment; unfortunately, this brings the risk of instability. The problem becomes more serious if the error sensors are far away from the secondary sources. As Berkhoff (2005) pointed out, the main objectives for the design of active sound barriers are: (1) no microphones in the far field should be used, but only microphones near the active sound barrier, (2) the system has to be suitable for moving noise sources, and (3) the system should consist of modules that work more or less independently by using a limited amount of information from other parts in the system in order to reduce communication bandwidth and hardware.

---

## 1.4 Virtual Sound Barriers

A virtual sound barrier (VSB) system is an array of acoustic sources and sensors forming an acoustic barrier, which blocks the direct propagation of sound without much blocking of air, light, and access (Qiu, Li, and Chen, 2005). It can create a quiet zone in a noisy environment by using the ANC-method, and the main mechanisms are absorption and/or reflection of the noise by the secondary sources. The theoretical background of the VSB can be traced back to the Huygens' principle, quantified in one way by the Kirchhoff-Helmholtz integral equation, which shows that, for a volume without internal sources, the sound pressure at any given location inside is completely determined by the sound pressure and its normal gradient on the boundary. Thus, if all sound pressure and its normal gradient on the boundary are reduced to zero, the sound pressure inside is zero too (Nelson and Elliott, 1992). In this section, the history, principle, and design methods of VSB systems are introduced briefly (Qiu and Zou, 2016). More specific knowledge on planar VSBs, three-dimensional VSBs, their applications, advantages, limitations, and the future research directions are presented in Chapters 2 to 5.

### 1.4.1 History

The idea of applying the Huygens' principle to active sound field control was pointed out by Jessel et al. in 1968, and then various numerical studies and several experiments were undertaken to demonstrate the feasibility of the idea; however, even now the technology is still far from a reality (Nelson and Elliott, 1992). In some research on active control of sound radiation from a primary source, such as a power transformer, a number of near-field secondary sources are used to control the total sound radiation, where the physical mechanism is to minimize the total power output of the primary and secondary sources rather than to control the sound field based on the Huygens principle as that used in the VSB systems (Mangiante, 1977).

The most closely related work to VSB systems was reported by Epain and Friot (2007), where 30 error microphones, 30 loudspeakers, and a 32-channel ANC controller were used to create a spherical quiet zone with a radius of 0.3 m by applying the boundary pressure control technique. The experiments in a quasi-anechoic environment show that the noise can be efficiently canceled everywhere inside the sphere over a wide frequency range for both pure tones and broadband noise. In the same year, the theoretical and experimental studies on a 16-channel cylindrical VSB system were conducted by Zou (2007) to investigate the feasibility of the VSB system. The experiments with this cylindrical VSB system in a normal room show that the average reduction of more than 10 dB inside a cylindrical region of 0.2 m height and 0.2 m radius can be achieved up to 550 Hz. The control performance of the system is affected by the distribution of the error sensors and the secondary sources, and the average noise reduction inside the target region decreases with the increase of the noise frequency at about −6 dB per 100 Hz (Zou et al., 2007).

Later, Zou and Qiu (2008) studied the effects on the performance of VSB systems of the presence of a human head, positioned within the quiet zone and surrounded by the error sensors. It is found that the introduction of the human head is beneficial to the VSB system in terms of performance robustness with regard to the movements of the human head. A comparison of three cost functions of the VSB system was undertaken with numerical simulations, and the best strategy was found to be minimizing the sum of the total acoustic energy density at the error sensors in terms of the total noise reduction and the uniformity degree of the sound attenuation distribution in the target region (Zou, Qiu, and Lu, 2009).

In practice, the noise reduction in a target region can achieve the maximum if the error sensors of the VSB systems are located at these target locations. However, this sort of arrangement results in the interference of the error sensors with the human head. This problem can be solved by using the virtual sensing strategy, in which the virtual sensors are placed in the target region while the physical sensors are placed at the border of the target region. It has been demonstrated that the introduction of the virtual sensors is feasible for developing a compact VSB system (Zou and Qiu, 2009).

Because a VSB system is usually used in a room, the system may be located near reflective surfaces. It is shown that when a VSB system is near a reflective surface, its noise reduction performance fluctuates periodically around the performance curve of the system without a nearby surface with the distance between the surface and the system. The performance of the VSB system is sensitive to the incident angle due to the reflective surface (Qiu, Zou, and Rao, 2009). The sound pressure in the exterior area of a VSB system usually increases due to the sound reflection mechanism of the system; however, the impact on exterior area can be reduced by using specially designed directional secondary sources (Rao, 2011). Some experiments have been carried out with a VSB system to control outdoor traffic noise along the highways (Zhao et al., 2017).

A planar VSB is a two-dimensional VSB system targeted to control sound propagation from one side of the plane. It can be used to control plane wave propagation in a free field (Guo and Pan, 1998a; Elliott et al., 2018), to control plane wave propagation through a finite aperture (Elliott et al., 2018; Lam et al., 2018), to control sound radiation from an opening of an enclosure (Wang, Tao, and Qiu, 2015, 2017; Wang et al., 2017, 2018a), and to control sound transmission via an opening into an enclosure (Lam et al., 2018; Wang, Tao, and Qiu, 2019). There are application examples of the planar VSBs for noise radiation control from power transformers in an enclosure (Tao, Wang, and Qiu, 2015; Tao et al., 2016) and for sound transmission control into a room through an open window (Murao et al., 2017; Lam et al., 2018b).

### 1.4.2 Principle

For a volume $V$ without internal sources, the sound pressure at any location inside is completely determined by the sound pressure and its normal gradient on the boundary, as described by the following Kirchhoff-Helmholtz equation (Nelson and Elliott, 1992)

$$p(\mathbf{r}) = \int_S \left[ G(\mathbf{r} \,|\, \mathbf{s}) \nabla p(\mathbf{s}) - p(\mathbf{s}) \nabla G(\mathbf{r} \,|\, \mathbf{s}) \right] \cdot \mathbf{n} \, dS \qquad (1.42)$$

where $V$ is surrounded by the surface $S$, $p(\mathbf{s})$ is the sound pressure at position $\mathbf{s}$ on the surface $S$, $p(\mathbf{r})$ is the sound pressure at position $\mathbf{r}$ inside the volume $V$, and $\mathbf{n}$ is the unit vector normal to the surface $S$ pointing outward from the volume. The free space Green function $G(\mathbf{r}|\mathbf{s})$ can be expressed as (Nelson and Elliott, 1992)

$$G(\mathbf{r}\,|\,\mathbf{s}) = \frac{1}{4\pi\,|\,\mathbf{r}-\mathbf{s}\,|}\,e^{-jk|\mathbf{r}-\mathbf{s}|} \tag{1.43}$$

The Kirchhoff–Helmholtz equation shows that the sound pressure inside the volume $V$ can be reduced by reducing the sound pressure and normal gradient on the boundary. This is the theoretical basis of the VSB systems. Figure 1.11 shows an example of a VSB system, in which an array of loudspeakers is located in a three-dimensional closed structure in order to create a quiet zone within the space surrounded by the error sensors in a noisy environment.

The monitoring sensors can be omitted for a steady primary noise field, where the radiation of the loudspeakers can be predesigned in advance according to the characteristics of the primary noise field, the location of the quiet zone and the arrangement of the secondary loudspeakers. For a general primary noise field where the position, the amplitude, and the frequency of the noise source are time-varying, good noise reduction performance of the VSB systems can be achieved with an adaptive controller that receives the input signals of the monitoring sensors and adjusts the output signals to the loudspeakers.

The physical mechanisms of the VSB system in Figure 1.11 might be different according to the type of secondary sources and control strategies. Since a VSB system is usually far away from the primary source, the mechanism is generally absorption or reflection of the sound energy of the primary source rather than reducing the impedance seen by the primary source (except for the VSB system inside an enclosure, which obtains control via strong modal coupling). For instance, the incident sound from the primary source to the quiet zone is reflected back when the sound pressure of the sensors

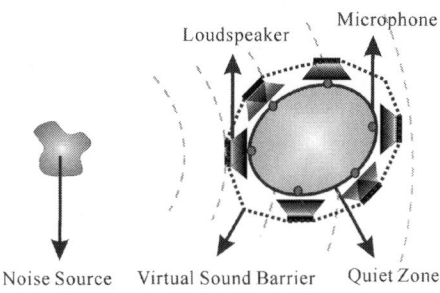

**FIGURE 1.11**
Schematic drawing of a VSB system.

is reduced to zero in Figure 1.11. As a result, the total sound energy in the environment is usually increased when the VSB system is working.

### 1.4.3 Design Methods

There are two main methods for designing a VSB system. One is the expansion method of the primary sound field, which is suitable for steady primary sound fields, and the other is the least mean square method, which is applicable to time-varying primary sound fields. A spherical VSB structure with a radius of $r_v$ is used in this section to illustrate the two methods. The origin of the coordinate system is set at the center of the spherical quiet area and the secondary sources are located on the surface of the sphere. There is no other sound source or sound scattering object inside the sphere.

#### 1.4.3.1 The Expansion Method of the Primary Sound Field

The primary sound field at point $\mathbf{r}=(r, \theta, \phi)$ can be expressed as (Williams, 1999)

$$p_p(r,\theta,\phi,k) = \begin{cases} \displaystyle\sum_{n=0}^{\infty}\sum_{m=-n}^{n} A_n^m(k)j_n(kr)Y_n^m(\theta,\phi), \; r < r_v \\[2em] \displaystyle\sum_{n=0}^{\infty}\sum_{m=-n}^{n} C_n^m(k)h_n(kr)Y_n^m(\theta,\phi), \; r > r_v \end{cases} \tag{1.44}$$

where $k$ is the wave number, $A_n^m(k)$ and $C_n^m(k)$ are the coefficients of the spherical harmonics $Y_n^m(\theta,\phi)$, $j_n(kr)$ is the spherical Bessel function of the first kind at the $n$th order, $h_n(kr)$ is the spherical Hankel function of the first kind at the $n$th order. $Y_n^m(\theta,\phi)$ is given by (Williams, 1999),

$$Y_n^m(\theta,\phi) = \sqrt{\frac{(2n+1)}{4\pi}\frac{(n-|m|)!}{(n+|m|)!}} P_n^{|m|}(\cos\theta)e^{-jm\phi} \tag{1.45}$$

where $P_n^{|m|}(\cdot)$ is the associated Legendre function of degree $n$ and order $m$.

The continuous monopole and dipole secondary sources are located on the surface of the spherical area. According to the Kirchhoff-Helmholtz equation, the secondary sound field inside the sphere can be represented by the sound pressure and its normal gradient on the boundary,

$$p_s(\mathbf{r}) = \int_0^{2\pi}\int_0^{\pi}\left[\frac{\partial p(\mathbf{r}_v)}{\partial n(\mathbf{r}_v)}G(\mathbf{r}\,|\,\mathbf{r}_v) - jkp(\mathbf{r}_v)\frac{1}{jk}\frac{\partial G(\mathbf{r}\,|\,\mathbf{r}_v)}{\partial n(\mathbf{r}_v)}\right]r_v^2\sin\theta_v\,d\theta_v d\phi_v \tag{1.46}$$

where $r_v = (r_v, \theta_v, \phi_v)$ is the coordinate of the point on the spherical surface. The source strengths of the monopole and dipole sources can be given by

$$S(\mathbf{r}_v) = \frac{\partial p(\mathbf{r}_v)}{\partial \mathbf{n}(\mathbf{r}_v)} \tag{1.47}$$

$$D(\mathbf{r}_v) = jkp(\mathbf{r}_v) \tag{1.48}$$

where the directivity of the dipole source is independent of the frequency due to the normalized coefficient $jk$ of the dipole source. The control objective is to reduce the total sound field inside the sphere to zero, so there is

$$p_t(\mathbf{r}) = p_p(\mathbf{r}) + p_s(\mathbf{r}) = 0, \quad |\mathbf{r}| \le r_v \tag{1.49}$$

If $p(r_v)$ in Equations (1.47) and (1.48) is set to $-p_p(r_v)$, zero sound pressure inside the quiet area can be achieved. Therefore, the source strengths of the secondary sources can be calculated by

$$S(\mathbf{r}_v) = -k \sum_{n=0}^{\infty} \sum_{m=-n}^{n} A_n^m(k) j_n'(kr_v) Y_n^m(\theta_v, \phi_v) \tag{1.50}$$

$$D(\mathbf{r}_v) = -jk \sum_{n=0}^{\infty} \sum_{m=-n}^{n} A_n^m(k) j_n(kr_v) Y_n^m(\theta_v, \phi_v) \tag{1.51}$$

The continuous sources have to be discretized in practice. With $N_s$ discrete sources, the secondary sound field is written as (Rao, 2011)

$$p_s(\mathbf{r}) = \sum_{l=1}^{N_s} r_v^2 \beta_l \left[ S(\mathbf{r}_l) G(\mathbf{r} \mid \mathbf{r}_l) - D(\mathbf{r}_l) \frac{1}{jk} \frac{\partial G(\mathbf{r} \mid \mathbf{r}_l)}{\partial \mathbf{n}(\mathbf{r}_l)} \right] \tag{1.52}$$

where $r_l = (r_v, \theta_l, \phi_l)$ is the coordinate of the $l$th source, $\beta_l$ is the weight coefficient of the spherical harmonics. When the order of the spherical harmonics expansion is chosen as $N$, the strengths of the discrete secondary sources are given by

$$S(\mathbf{r}_l) = -k \sum_{n=0}^{N} \sum_{m=-n}^{n} A_n^m(k) j_n'(kr_v) Y_n^m(\theta_l, \phi_l) \tag{1.53}$$

$$D(\mathbf{r}_l) = -jk \sum_{n=0}^{N} \sum_{m=-n}^{n} A_n^m(k) j_n(kr_v) Y_n^m(\theta_l, \phi_l) \tag{1.54}$$

### 1.4.3.2 The Least Mean Square Method

Assume $N_s$ first-order secondary sources of a VSB system are uniformly distributed on the spherical surface with a radius of $r_v$ at the coordinate $\mathbf{r}_l = (r_v, \theta_l, \phi_l)$. The secondary sound field can be expressed as (Rao, 2011)

$$p_s(\mathbf{r}) = \sum_{l=1}^{N_s} \frac{q_l e^{-jk|\mathbf{r}-\mathbf{r}_l|}}{4\pi|\mathbf{r}-\mathbf{r}_l|} \left[ a - (1-a)\left(1 - \frac{j}{k|\mathbf{r}-\mathbf{r}_l|}\right)\cos\gamma \right] \qquad (1.55)$$

where $q_l$ is the strength of the $l$th secondary source. The part in the square brackets on the right-hand side of the equation is the general expression of the first-order directional source, and the directivity coefficient $\alpha$ is between 0 and 1 (Williams, 1999). When $\alpha$ is equal to 1, 0, and 0.5, the source is a monopole (omnidirectional source), a dipole (figure-eight source) and a tripole (hypercardioid source), respectively. $\gamma$ is the angle between the vector $\mathbf{r} - \mathbf{r}_l$ and the axis of the source.

The total sound field with control is given by

$$p_t(\mathbf{r}) = p_p(\mathbf{r}) + p_s(\mathbf{r}) \qquad (1.56)$$

The cost functions to be minimized can be the sum of acoustic potential energy density, the sum of acoustic kinetic energy density, and the sum of the total acoustic energy density at the error sensors (Zou, 2007),

$$J_p = \sum_{i=1}^{N_e} |p_t(\mathbf{r}_{e,i})|^2 + \beta \mathbf{q}_s^H \mathbf{q}_s \qquad (1.57)$$

$$J_k = \sum_{i=1}^{N_e} |v_t(\mathbf{r}_{e,i})|^2 + \beta \mathbf{q}_s^H \mathbf{q}_s \qquad (1.58)$$

$$J_e = \sum_{i=1}^{N_e} \left[ \frac{1}{2\rho_0 c_0^2} |p_t(\mathbf{r}_{e,i})|^2 + \frac{\rho_0}{2} |v_t(\mathbf{r}_{e,i})|^2 \right] + \beta \mathbf{q}_s^H \mathbf{q}_s \qquad (1.59)$$

where $c_0$ is the speed of sound in the air, $\rho_0$ is the air density, $N_e$ is the number of error sensors located at $\{\mathbf{r}_{e,i}, i = 1, 2, \ldots, N_e\}$, $\mathbf{q}_s = [q_1, q_2, \ldots, q_{Ns}]^T$ is the vector of the secondary source strengths, $v_t(r)$ is the normal component of the particle velocity. $\beta$ can be set to an appropriate value to constrain the secondary source output power. The error sensors are usually located at the boundary of the quiet zone with a radius smaller than $r_v$.

The cost functions in Equations (1.57–1.59) can be expressed in the quadratic form as (Zou, 2007)

$$J = \mathbf{q}_s^H (\mathbf{A} + \beta I)\mathbf{q}_s + \mathbf{q}_s^H \mathbf{b} + \mathbf{b}^H \mathbf{q}_s + c \qquad (1.60)$$

where **A** is a matrix related to the transfer functions between the pressure (and/or particle velocity) at error sensors and the secondary source strengths, **b** is a vector related to the transfer functions mentioned above and the primary sound field, $c$ is a constant only related to the primary sound field. For example, when the cost function is the acoustic potential energy density shown in Equation (1.57), the corresponding parameters can be expressed as (Zou, 2007)

$$\mathbf{A} = \mathbf{Z}_s^H \mathbf{Z}_s, \quad \mathbf{b} = \mathbf{Z}_s^H \mathbf{p}_p, \quad c = \mathbf{p}_p^H \mathbf{p}_p, \tag{1.61}$$

$$\mathbf{Z}_s = \begin{bmatrix} Z_{se}(\mathbf{r}_{e,1}|\mathbf{r}_{s,1}) & Z_{se}(\mathbf{r}_{e,1}|\mathbf{r}_{s,2}) & \cdots & Z_{se}(\mathbf{r}_{e,1}|\mathbf{r}_{s,N_s}) \\ Z_{se}(\mathbf{r}_{e,2}|\mathbf{r}_{s,1}) & Z_{se}(\mathbf{r}_{e,2}|\mathbf{r}_{s,2}) & \cdots & Z_{se}(\mathbf{r}_{e,2}|\mathbf{r}_{s,N_s}) \\ \cdots & \cdots & \cdots & \cdots \\ Z_{se}(\mathbf{r}_{e,N_e}|\mathbf{r}_{s,1}) & Z_{se}(\mathbf{r}_{e,N_e}|\mathbf{r}_{s,2}) & \cdots & Z_{se}(\mathbf{r}_{e,N_e}|\mathbf{r}_{s,N_s}) \end{bmatrix} \tag{1.62}$$

$$\mathbf{p}_p = \begin{bmatrix} p_p(\mathbf{r}_{e,1}) & p_p(\mathbf{r}_{e,2}) & \cdots & p_p(\mathbf{r}_{e,N_e}) \end{bmatrix}^T \tag{1.63}$$

where

$$Z_{se}(\mathbf{r}_{e,i}|\mathbf{r}_{s,j}) = \frac{e^{-jk|\mathbf{r}_{e,i}-\mathbf{r}_{s,j}|}}{4\pi|\mathbf{r}_{e,i}-\mathbf{r}_{s,j}|} \left[ a - (1-a)\left(1 - \frac{j}{k|\mathbf{r}_{e,i}-\mathbf{r}_{s,j}|}\right) \cos\gamma_{i,j} \right] \tag{1.64}$$

$$\cos\gamma_{i,j} = \frac{(\mathbf{r}_{e,i}-\mathbf{r}_{s,j})\cdot\mathbf{r}_{s,j}}{|\mathbf{r}_{e,i}-\mathbf{r}_{s,j}||\mathbf{r}_{s,j}|} \tag{1.65}$$

The optimal strength of the secondary sources for all the above cost functions is given by (Zou, 2007)

$$\mathbf{q}_s = -(\mathbf{A}+\beta I)^{-1}\mathbf{b} \tag{1.66}$$

In the following chapters of the book, two particular VSB systems are discussed. The first is the planar VSB system and its control on wave propagation in a free field, on wave propagation through a finite aperture, on sound radiation from an opening of an enclosure, and on sound transmission via an opening into an enclosure. The second is the three-dimensional VSB system, which is used to create a quiet zone in a noisy environment. The effects of a diffracting sphere inside the quiet zone, a reflective surface near the system, and the cost functions for optimizing the system are discussed.

# 2

## Planar Virtual Sound Barriers

### 2.1 Problem Description

Planar virtual sound barriers are used to control sound propagation from one side of a plane. Figure 2.1 shows an example of a planar virtual sound barrier with an array of secondary sources. The primary sound field can be plane waves, or waves generated by point monopole sources, or a combination of the two. The questions to be answered are: Is it feasible to control sound propagation with the planar virtual sound barrier? What is the requirement for the space between the secondary sources? Are there any applications with such a planar virtual sound barrier? And how to implement it in practice?

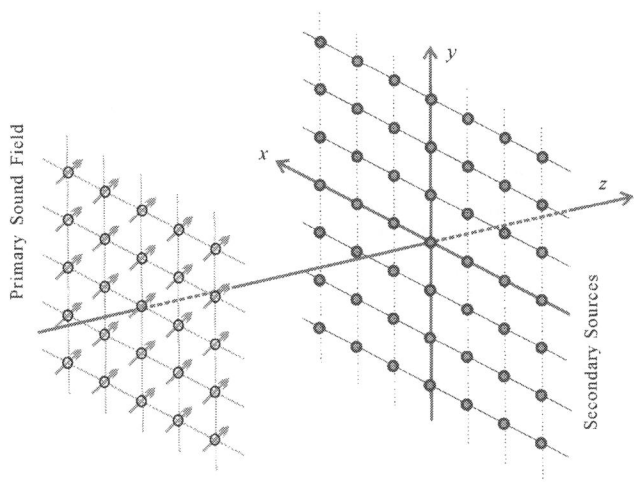

**FIGURE 2.1**
An example of a planar virtual sound barrier with an array of secondary point monopole sources on a plane.

The virtual sound barriers are usually far away from the primary sound sources, so the secondary sources can only interact with the primary sound field and have little effect on the primary sources. This implies the sound control mechanisms of the virtual sound barriers are not like those in the active sound radiation control, where the secondary sources are usually located close to the primary sources in order to reduce the effective "loading" on the primary sound sources, with the result that the total sound power radiated from the whole system is reduced (Nelson and Elliott, 1992). The mechanisms for the planar virtual sound barriers to control sound propagation are reflection or absorption, or a combination of the two.

## 2.2 Control of Sound Propagation in Free Fields

This section discusses the control of sound propagation in free fields. Two types of primary sound fields are considered, one is that from plane waves, and the other is the sound field generated by point monopole sources.

### 2.2.1 Control of the Plane Wave Primary Sound Field

Assume that the primary sound field is that from a plane wave with an amplitude of $A_p$, so the sound pressure at location $\mathbf{r} = (x, y, z)$ can be expressed by

$$p_p(\mathbf{r}) = A_p e^{-jk\mathbf{n}\cdot\mathbf{r}} \tag{2.1}$$

where $\mathbf{n} = (n_x, n_y, n_z)$ is the unit vector in the propagation direction of the plane wave, $k = \omega/c_0$ is the wavenumber, $c_0$ is the speed of sound in the air, $\omega = 2\pi f$ is the angular frequency, and $f$ is the frequency. The secondary sound field generated by the planar virtual sound barrier is the superposition of many secondary sources. Assume the secondary sources are point monopoles, the secondary sound field can be expressed by

$$p_s(\mathbf{r}) = \sum_{i=1}^{\infty} \frac{j\omega\rho_0 q_i}{4\pi|\mathbf{r} - \mathbf{s}_i|} e^{-jk|\mathbf{r} - \mathbf{s}_i|} \tag{2.2}$$

where $\rho_0$ is the air density, and $q_i$ is the complex source strength of the $i$th secondary source located at $s_i$. The total sound pressure at location $\mathbf{r}$ can be calculated by

$$p_t(\mathbf{r}) = A_p e^{-jk\mathbf{n}\cdot\mathbf{r}} + \sum_{i=1}^{\infty} \frac{j\omega\rho_0 q_i}{4\pi|\mathbf{r} - \mathbf{s}_i|} e^{-jk|\mathbf{r} - \mathbf{s}_i|} \tag{2.3}$$

The particle velocity of the total sound in the propagating z direction at location $\mathbf{r}$ on the error sensor plane can be calculated by

$$v_z(\mathbf{r}) = \frac{A_p k_z}{k \rho_0 c_0} e^{-jk\mathbf{n}\cdot\mathbf{r}} + \sum_{i=1}^{\infty} \frac{jkq_i d_{se}}{4\pi|\mathbf{r}-\mathbf{s}_i|^2}\left(1 - \frac{j}{k|\mathbf{r}-\mathbf{s}_i|}\right)e^{-jk|\mathbf{r}-\mathbf{s}_i|} \quad (2.4)$$

where $d_{se}$ is the distance between the virtual sound barrier plane and the error sensor plane. The mean active sound intensity in the z direction at location $\mathbf{r}$ can be calculated by

$$I_z(\mathbf{r}) = \frac{1}{2}\text{Re}\{p_t^*(\mathbf{r})v_z(\mathbf{r})\} \quad (2.5)$$

where Re{·} denotes the real part of the complex quantity in the curly brackets and * denotes complex conjugation.

The objective of the planar virtual sound barrier is to reduce the sound power propagation in the z direction behind the planar virtual sound barrier, which is defined as

$$J(d_{se}) = \int_{-\infty}^{\infty}\int_{-\infty}^{\infty} I_z(x,y,d_{se})dxdy \quad (2.6)$$

Substituting Equations (2.3) and (2.4) into Equations (2.5) and (2.6), the cost function can be expressed as a function of the complex source strength, whose optimal value can be obtained by minimizing the cost function. Unfortunately, it is hard to derive an analytical solution for a problem with an infinite number of secondary sources.

One way of deriving an analytical solution is to use wavenumber domain analyses (Elliott et al., 2018). The primary sound field in Equation (2.1) can be written as

$$p_p(x,y,z) = A_p e^{-jk(x\sin\theta\cos\phi+y\sin\theta\sin\phi+z\cos\theta)} \quad (2.7)$$

where $\theta$ and $\phi$ are the propagation angles of the plane wave. The associated particle velocity in the z direction is

$$v_p(x,y,z) = \frac{A_p\cos\theta}{\rho_0 c_0} e^{-jk(x\sin\theta\cos\phi+y\sin\theta\sin\phi+z\cos\theta)} \quad (2.8)$$

To be able to cancel the primary sound field propagation effectively, the strength distribution of the secondary source array should be

$$v_s(x,y,0) = \sum_{m_1=-\infty}^{\infty}\sum_{m_2=-\infty}^{\infty} q_{m_1,m_2}\delta(x-m_1d)\delta(y-m_2d)e^{-jk(x\sin\theta\cos\phi+y\sin\theta\sin\phi)} \quad (2.9)$$

where $\delta(\cdot)$ is the Dirac delta function, $m_1$ and $m_2$ are the indexes of the secondary sources, and $d$ is the separation distance between the secondary sources. For simplicity of the discussion, assume that the primary sound field is generated by a normally incident plane wave, so $\theta = 0$, then the strength of all the secondary sources should be the same. Equation (2.9) is simplified to,

$$v_s(x, y, 0) = \sum_{m_1=-\infty}^{\infty} \sum_{m_2=-\infty}^{\infty} q_s \delta(x - m_1 d)\delta(y - m_2 d) \tag{2.10}$$

where $q_s$ is the strength of one secondary source. Expanding this periodic function with the Fourier series gives

$$v_s(x, y, 0) = \sum_{n_1=-\infty}^{\infty} \sum_{n_2=-\infty}^{\infty} c_{n_1,n_2} e^{-j\frac{2\pi}{d}(n_1 x + n_2 y)} \tag{2.11}$$

where

$$c_{n_1,n_2} = \frac{1}{d^2} \int_0^d \int_0^d \sum_{m_1=-\infty}^{\infty} \sum_{m_2=-\infty}^{\infty} q_s \delta(x - m_1 d)\delta(y - m_2 d) e^{j\frac{2\pi}{d}(n_1 x + n_2 y)} dx dy \tag{2.12}$$

Define $k_{x,\,n1} = 2\pi n_1/d$, $k_{y,\,n2} = 2\pi n_2/d$, Equation (2.11) can be rewritten as

$$v_s(x, y, 0) = V_s \sum_{n_1=0}^{\infty} \sum_{n_2=0}^{\infty} \varepsilon_{n_1} \varepsilon_{n_2} \cos(k_{x,n_1} x)\cos(k_{x,n_2} y) \tag{2.13}$$

where $V_s = q_s/d^2$, $\varepsilon_n = 1$ for $n = 0$ and $\varepsilon_n = 2$ for $n > 0$.

This velocity distribution propagates in the $z$ direction, and the particle velocity of the generated sound in the $z$ direction at location $\mathbf{r} = (x, y, z)$ can be written as (Williams, 1999)

$$v_z(\mathbf{r}) = V_s \sum_{n_1=0}^{\infty} \sum_{n_2=0}^{\infty} \varepsilon_{n_1} \varepsilon_{n_2} \cos(k_{x,n_1} x)\cos(k_{x,n_2} y) e^{-jk_{n_1,n_2} z} \tag{2.14}$$

where the wavenumber in the $z$ direction for the spatial harmonic with an order of $n_1$ and $n_2$ is

$$k_{n_1,n_2} = \sqrt{k^2 - (k_{x,n_1}^2 + k_{y,n_2}^2)} \tag{2.15}$$

If $k < \sqrt{k_{x,n_1}^2 + k_{y,n_2}^2}$, then $k_{n_1,n_2}$ is imaginary and the terms are evanescent, so the components of the wave corresponding to these terms cannot propagate to the far field. Because the primary sound field is generated by a normally incident plane wave, so

$$v_p(x, y, z) = \frac{A_p}{\rho_0 c_0} e^{-jkz} \tag{2.16}$$

Therefore, the particle velocity of the total sound in the $z$ direction at the location due to both the primary and secondary sources is

$$v_z(x,y,z) = \frac{A_p}{\rho_0 c_0} e^{-jkz} + V_s \sum_{n_1=0}^{\infty} \sum_{n_2=0}^{\infty} \varepsilon_{n_1} \varepsilon_{n_2} \cos(k_{x,n_1}x)\cos(k_{x,n_2}y)e^{-jk_{n_1,n_2}z} \quad (2.17)$$

The pressure associated with the particle velocity distribution can be written as

$$p(x,y,z) = (A_p + \rho_0 c_0 V_s)e^{-jkz}$$
$$+\rho_0 c_0 V_s \sum_{n_1+n_2=1}^{\infty} \frac{k\varepsilon_{n_1}\varepsilon_{n_2}}{k_{n_1,n_2}} \cos(k_{x,n_1}x)\cos(k_{x,n_2}y)e^{-jk_{n_1,n_2}z} \quad (2.18)$$

where the summation from $n_1 + n_2 = 1$ to $\infty$ means $n_1$ and $n_2$ can be any values from 0 to $\infty$ except that $n_1 + n_2 = 0$. The spatially averaged intensity propagating in the $z$ direction behind the planar virtual sound barrier can be calculated with

$$\Pi(z) = \frac{1}{2d^2} \text{Re}\left\{ \int_0^d \int_0^d p^*(x,y,z)v_z(x,y,z)dxdy \right\} \quad (2.19)$$

Substituting Equations (2.17) and (2.18) into it yields,

$$\Pi(z) = \frac{1}{2}\left[ \rho_0 c_0 \left( 1 + \sum_{n_1+n_2=1}^{\infty} \varepsilon_{n_1}\varepsilon_{n_2} \text{Re}\left\{ \frac{k}{k_{n_1,n_2}} e^{-j(k_{n_1,n_2}-k^*_{n_1,n_2})z} \right\} \right) V_s^2 \right.$$
$$\left. + A_p^* V_s + A_p V_s^* + \frac{A_p^2}{\rho_0 c_0} \right] \quad (2.20)$$

If $k < \sqrt{k_{x,m}^2 + k_{y,n_2}^2}$, then $k_{n_1,n_2}$ is imaginary, the term in Re$\{\}$ is a pure imaginary, so the term with Re$\{\}$ is zero. This indicates that high-order spatial harmonic modes do not contribute to the intensity propagating in the $z$ direction behind the planar virtual sound barrier in the low-frequency range, because they are evanescent waves and cannot propagate to the far field. Assume $N_1$ and $N_2$ are the largest numbers for $n_1$ and $n_2$ that make $k_{n_1,n_2}$ real, then Equation (2.20) changes to

$$\Pi(z) = \frac{1}{2}\left[ \rho_0 c_0 \left( 1 + \sum_{n_1+n_2=1}^{N_1,N_2} \varepsilon_{n_1}\varepsilon_{n_2}\frac{k}{k_{n_1,n_2}} \right) V_s^2 + A_p^* V_s + A_p V_s^* + \frac{A_p^2}{\rho_0 c_0} \right] \quad (2.21)$$

This cost function is minimized if the secondary source strength

$$V_s = -\frac{A_p}{\rho_0 c_0 \left( 1 + \sum_{n_1+n_2=1}^{N_1,N_2} \varepsilon_{n_1} \varepsilon_{n_2} \frac{k}{k_{n_1,n_2}} \right)} \tag{2.22}$$

in which case, the minimum of the spatially averaged intensity propagating in the $z$ direction behind the planar virtual sound barrier is

$$\Pi_{min} = \frac{1}{2} \frac{A_p^2}{\rho_0 c_0} \left( \frac{\sum_{n_1+n_2=1}^{N_1,N_2} \varepsilon_{n_1} \varepsilon_{n_2} \frac{k}{k_{n_1,n_2}}}{1 + \sum_{n_1+n_2=1}^{N_1,N_2} \varepsilon_{n_1} \varepsilon_{n_2} \frac{k}{k_{n_1,n_2}}} \right) \tag{2.23}$$

When there is no secondary source, the primary spatially averaged intensity propagating in the $z$ direction behind the planar virtual sound barrier is

$$\Pi_0 = \frac{1}{2} \frac{A_p^2}{\rho_0 c_0} \tag{2.24}$$

Therefore, the attenuation in the spatially averaged intensity propagating in the $z$ direction is

$$\text{Att} = 10 \log_{10} \frac{\Pi_0}{\Pi_{min}} = 10 \log_{10} \left( \frac{1 + \sum_{n_1+n_2=1}^{N_1,N_2} \varepsilon_{n_1} \varepsilon_{n_2} \frac{k}{k_{n_1,n_2}}}{\sum_{n_1+n_2=1}^{N_1,N_2} \varepsilon_{n_1} \varepsilon_{n_2} \frac{k}{k_{n_1,n_2}}} \right) \tag{2.25}$$

The attenuation is infinitely large at low frequency when only $k_{0,0}$ is real. In this case, all higher-order spatial harmonic modes generated by the secondary sources are evanescent waves and cannot propagate to the far field, so the primary sound field can be canceled completely with the secondary sound field without any anti-facts (introducing extra secondary sound in the far field). At higher frequencies, when the wavenumbers of the other modes other than the (0, 0)th mode are real, the secondary sound field generated by the array of secondary sources consists not only of the plane wave in the normal direction but also of the waves propagating in other directions, which are the anti-facts for the sound attenuation. Thus, there is a balance between the degree of canceling the primary sound and generating the extra sound. At very high frequencies, the best way is not to generate any secondary sound, but there is little attenuation either.

Figure 2.2 shows the normalized real parts of the wavenumbers of the spatial harmonic with orders of (0, 0), (0, 1), (1, 1), (0, 2), and (1, 2) (curves from top to bottom) and the attenuation as a function of normalized source separation distance $d/\lambda$ (where $\lambda$ is the wavelength). The normalized distance is proportional to frequency $f$ by $d/\lambda = df/c_0$. The normalized real part of the wavenumbers of the $(n_1, n_2)$th spatial harmonic is calculated by

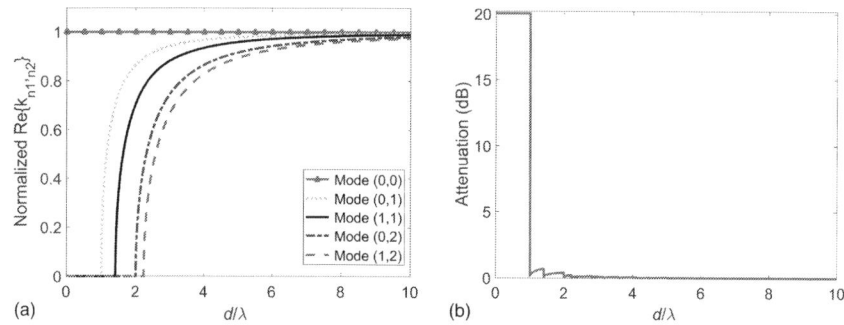

**FIGURE 2.2**
(a) The normalized real parts of the wavenumbers of the spatial harmonic with orders of $(0, 0)$, $(0, 1)$, $(1, 1)$, $(0, 2)$, and $(1, 2)$ (curves from top to bottom), (b) the attenuation of the planar virtual sound barrier system as a function of the normalized source separation distance $d/\lambda$.

$$\frac{k_{n_1,n_2}}{k} = \sqrt{1 - \frac{(n_1^2 + n_2^2)}{(d/\lambda)^2}} \qquad (2.26)$$

In Figure 2.2(a), at low frequency when $d < \lambda$, only the wavenumber of the $(0, 0)$th mode is real and it has $k_{0,0}/k = 1$. With the frequency increasing, more and more modes have real wavenumbers, so they can propagate to the far field. In Figure 2.2(b), at low frequency when $d < \lambda$, the attenuation is infinitely large (clipped to 20 dB in the figure). At the frequency that has $d = \lambda$, the attenuation drops to zero, because the secondary source strength has to be zero due to the counteraction between them. At higher frequencies, there is some attenuation, but very little, for example, the maximum is less than 1 dB.

It should be noted that the source separation distance requirement for effective attenuation under oblique incidence depends on the incidence angle and needs to be smaller than that in the normal incidence case for similar attenuation (Elliott et al., 2018). For example, for the grazing incidence plane wave with an incidence angle of nearly 90°, the attenuation is infinitely large only when $d < \lambda/2$.

The mechanism used for sound attenuation behind the planar virtual sound barrier with an array of monopole sources is acoustic wave reflection, which reflects the primary incidence sound energy back in the coming of the incidence wave. Therefore, a standing wave can be formed between the forward-going incident primary plane wave and the backward-going plane wave generated by the secondary source array (Elliott et al., 2018). Other mechanisms can also be adopted to reduce sound propagation behind the virtual sound barrier. For example, a combination of monopoles and dipoles can be used as the secondary sources to absorb the incident wave, which attenuates the sound propagation behind the virtual sound barrier but has little effect on the sound field in front of it (Elliott et al., 2018; Nelson and Elliott, 1992). The cost function to be minimized then becomes the sound power output of each secondary source. For example, the sound power output of a monopole with source strength of $q_s$ located at $\mathbf{r}_s = (x_s, y_s, z_s)$ is (Nelson and Elliott, 1992),

$$W_s(\mathbf{r}_s) = \frac{1}{2} \mathrm{Re} \left\{ \left[ p_p(\mathbf{r}_s) + p_s(\mathbf{r}_s) \right]^* q_s \right\} \tag{2.27}$$

where $p_s(\mathbf{r}_s)$ is the pressure produced by the secondary source upon itself, and $p_p(\mathbf{r}_s)$ is the pressure of the primary sound. After minimizing the cost function, the optimized source strength can be obtained, which is 180° out of phase with the primary sound pressure. The corresponding minimum power output has a negative value, which is the maximum power absorbed by the monopole. The absorption area is given by $\lambda^2/4\pi$, which is a circle of one wavelength in circumference. By using a combination of monopoles and dipoles, the equivalent sound absorption area can be increased by four times to $\lambda^2/\pi$. When the secondary sources are arranged in an array, the neighboring sources might affect the absorption of the current source, so the absorption area could be different. A general observation is that if the secondary sources are spaced at intervals of approximate $\lambda/2$, the absorption areas of each individual secondary sources can be merged into a large extended area with substantial absorption, so large sound attenuation behind the virtual sound barrier can be expected (Nelson and Elliott, 1992).

### 2.2.2 Control of the Primary Sound Field Generated by Monopole Sources

In practice, it is very hard to have a plane wave primary sound field, so the primary sound field generated by point monopole sources is considered in this section. Figure 2.3 shows an example of a planar virtual sound barrier with an array of secondary sources to control the primary sound field generated by a monopole source, where the distance between the primary source and the center of the secondary source array is $d_{ps}$, the distance between the secondary source plane and the error sensor plane is $d_{se}$, the separation distance between the secondary sources is $d_{ss}$, and the separation distance between the error sensors is $d_{ee}$ (Guo and Pan, 1999).

Assume that the primary sound pressure at location $\mathbf{r} = (x, y, z)$ is generated by $N_p$ point monopole sources by

$$p_p(\mathbf{r}) = \sum_{i=1}^{N_p} \frac{j\omega\rho_0 q_{p,i}}{4\pi |\mathbf{r} - \mathbf{r}_{p,i}|} e^{-jk|\mathbf{r} - \mathbf{r}_{p,i}|} \tag{2.28}$$

where $\mathbf{r}_{p,i} = (x_{p,i}, y_{p,i}, z_{p,i})$ is the location of the $i$th primary source with a complex strength of $q_{p,i}$, $\rho_0$ is the air density, $\omega = 2\pi f$ is the angular frequency, $f$ is the frequency, $k = \omega/c_0$ is the wavenumber, and $c_0$ is the speed of sound in the air. The secondary sound field generated by the planar virtual sound barrier

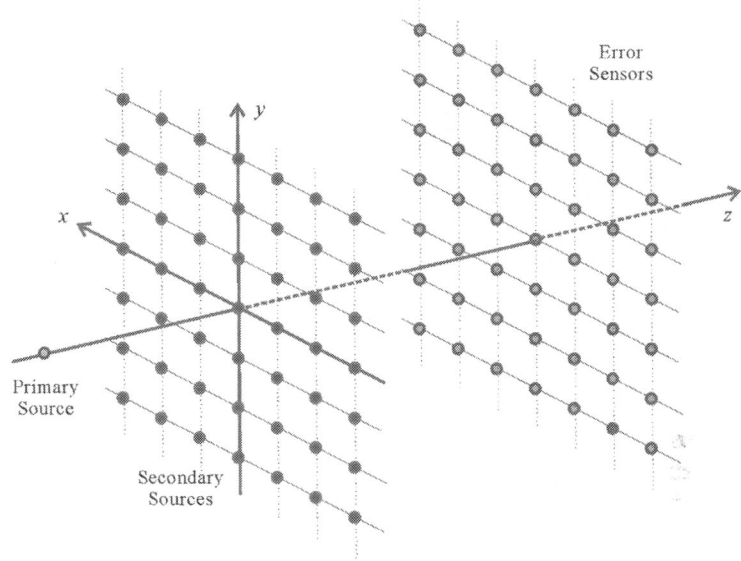

**FIGURE 2.3**
A planar virtual sound barrier with an array of secondary sources to control the primary sound field generated by a point monopole source.

is the superposition of $N_s$ secondary sources. Assume the secondary sources are point monopoles, the secondary sound field can be expressed by

$$p_s(\mathbf{r}) = \sum_{i=1}^{N_s} \frac{j\omega\rho_0 q_{s,i}}{4\pi|\mathbf{r}-\mathbf{r}_{s,i}|} e^{-jk|\mathbf{r}-\mathbf{r}_{s,i}|} \tag{2.29}$$

where $q_{s,i}$ is the complex source strength of the $i$th secondary source located at $\mathbf{r}_{s,i}$. The total sound pressure at location $\mathbf{r}$ can be calculated by

$$p_t(\mathbf{r}) = \sum_{i=1}^{N_p} \frac{j\omega\rho_0 q_{p,i}}{4\pi|\mathbf{r}-\mathbf{r}_{p,i}|} e^{-jk|\mathbf{r}-\mathbf{r}_{p,i}|} + \sum_{i=1}^{N_s} \frac{j\omega\rho_0 q_{s,i}}{4\pi|\mathbf{r}-\mathbf{r}_{s,i}|} e^{-jk|\mathbf{r}-\mathbf{r}_{s,i}|} \tag{2.30}$$

In matrix form, it has

$$p_t(\mathbf{r}) = \mathbf{Z}_p(\mathbf{r})\mathbf{q}_p + \mathbf{Z}_s(\mathbf{r})\mathbf{q}_s \tag{2.31}$$

with $\mathbf{q}_p = [q_{p,1}, q_{p,2}, \ldots, q_{p,Np}]^T$, $\mathbf{q}_s = [q_{s,1}, q_{s,2}, \ldots, q_{s,Ns}]^T$, $\mathbf{Z}_p(\mathbf{r}) = [Z_{p,1}(\mathbf{r}), Z_{p,2}(\mathbf{r}), \ldots, Z_{p,Np}(\mathbf{r})]$, $\mathbf{Z}_s(\mathbf{r}) = [Z_{s,1}(\mathbf{r}), Z_{s,2}(\mathbf{r}), \ldots, Z_{s,Ns}(\mathbf{r})]$, and the $i$th elements of $\mathbf{Z}_p(\mathbf{r})$ and $\mathbf{Z}_s(\mathbf{r})$ are respectively,

$$Z_{p,i}(\mathbf{r}) = \frac{j\omega\rho_0}{4\pi|\mathbf{r}-\mathbf{r}_{p,i}|} e^{-jk|\mathbf{r}-\mathbf{r}_{p,i}|}, \quad Z_{s,i}(\mathbf{r}) = \frac{j\omega\rho_0}{4\pi|\mathbf{r}-\mathbf{r}_{s,i}|} e^{-jk|\mathbf{r}-\mathbf{r}_{s,i}|} \tag{2.32}$$

The cost function to be minimized is defined as,

$$J = \mathbf{p}_t^H \mathbf{p}_t + \beta \mathbf{q}_s^H \mathbf{q}_s \tag{2.33}$$

where $\mathbf{p}_t = [p_t(\mathbf{r}_{e,1}), p_t(\mathbf{r}_{e,2}), \ldots, p_t(\mathbf{r}_{e,Ne})]^T$, $N_e$ is the number of error sensors located at $\mathbf{r}_{e,i}$, and $\beta$ is a positive real number being used to determine the weighting for the control effort term. Substituting Equation (2.31) into Equation (2.33), it has

$$J = \mathbf{q}_s^H (\mathbf{A} + \beta\mathbf{I})\mathbf{q}_s + \mathbf{q}_s^H \mathbf{b} + \mathbf{b}^H \mathbf{q}_s + c \tag{2.34}$$

with

$$\mathbf{A} = \mathbf{Z}_{se}^H \mathbf{Z}_{se}, \quad \mathbf{b} = \mathbf{Z}_{se}^H \mathbf{p}_p, \quad c = \mathbf{p}_p^H \mathbf{p}_p \tag{2.35}$$

and $\mathbf{Z}_{se} = [\mathbf{Z}_s(\mathbf{r}_{e,1})^T, \mathbf{Z}_s(\mathbf{r}_{e,2})^T, \ldots, \mathbf{Z}_s(\mathbf{r}_{e,Ne})^T]^T$ and $\mathbf{p}_p = [p_p(\mathbf{r}_{e,1}), p_p(\mathbf{r}_{e,2}), \ldots, p_p(\mathbf{r}_{e,Ne})]^T$. The optimal vector of the secondary source strength is given by

$$\mathbf{q}_s = -(\mathbf{A} + \beta\mathbf{I})^{-1}\mathbf{b} \tag{2.36}$$

After obtaining the optimal secondary source strength, it can be substituted back to Equation (2.30) to calculate the total sound pressure amplitude with control, which is denoted as $p_{t,o}$. The performance of the virtual sound barrier system is defined as the ratio of the sum of the squared sound pressure in the target area without and with control as

$$NR = 10\log_{10} \frac{\sum\limits_{i=1}^{N_v} |p_p(\mathbf{r}_{v,i})|^2}{\sum\limits_{i=1}^{N_v} |p_{t,o}(\mathbf{r}_{v,i})|^2} \tag{2.37}$$

where $\mathbf{r}_{v,i}$, $i = 1, 2, \ldots, N_v$, are the locations of the evaluation points, and $N_v$ is the number of evaluation points, which is chosen to ensure at least six evaluation points per wavelength.

The total sound power output of the control system (including both the primary and secondary sources) can be calculated with (Guo and Pan, 1999)

$$W_T = \frac{1}{2}\text{Re}\left\{\mathbf{q}_p^H \left(\mathbf{Z}_{pp}\mathbf{q}_p + \mathbf{Z}_{sp}\mathbf{q}_s\right)\right\} + \frac{1}{2}\text{Re}\left\{\mathbf{q}_s^H \left(\mathbf{Z}_{ps}\mathbf{q}_p + \mathbf{Z}_{ss}\mathbf{q}_s\right)\right\} \tag{2.38}$$

where $\mathbf{Z}_{pp} = [\mathbf{Z}_p(\mathbf{r}_{p,1})^T, \mathbf{Z}_p(\mathbf{r}_{p,2})^T, \ldots, \mathbf{Z}_p(\mathbf{r}_{p,Np})^T]^T$, $\mathbf{Z}_{sp} = [\mathbf{Z}_s(\mathbf{r}_{p,1})^T, \mathbf{Z}_s(\mathbf{r}_{p,2})^T, \ldots, \mathbf{Z}_s(\mathbf{r}_{p,Np})^T]^T$, $\mathbf{Z}_{ps} = [\mathbf{Z}_p(\mathbf{r}_{s,1})^T, \mathbf{Z}_p(\mathbf{r}_{s,2})^T, \ldots, \mathbf{Z}_p(\mathbf{r}_{s,Ns})^T]^T$, and $\mathbf{Z}_{ss} = [\mathbf{Z}_s(\mathbf{r}_{s,1})^T, \mathbf{Z}_s(\mathbf{r}_{s,2})^T, \ldots, \mathbf{Z}_s(\mathbf{r}_{s,Ns})^T]^T$. The real operation can be moved into the impedance terms (Nelson and Elliott, 1992, p. 270), so the above equation becomes

$$W_{\mathrm{T}} = \frac{1}{2}\left[\mathbf{q}_{\mathrm{p}}^{\mathrm{H}}\operatorname{Re}\{\mathbf{Z}_{\mathrm{pp}}\}\mathbf{q}_{\mathrm{p}} + \mathbf{q}_{\mathrm{p}}^{\mathrm{H}}\operatorname{Re}\{\mathbf{Z}_{\mathrm{sp}}\}\mathbf{q}_{\mathrm{s}} + \mathbf{q}_{\mathrm{s}}^{\mathrm{H}}\operatorname{Re}\{\mathbf{Z}_{\mathrm{ps}}\}\mathbf{q}_{\mathrm{p}} + \mathbf{q}_{\mathrm{s}}^{\mathrm{H}}\operatorname{Re}\{\mathbf{Z}_{\mathrm{ss}}\}\mathbf{q}_{\mathrm{s}}\right] \quad (2.39)$$

When the secondary sources are far from the primary source, $|\mathbf{Z}_{\mathrm{sp}}|$ and $|\mathbf{Z}_{\mathrm{ps}}|$ would be much smaller than $|\mathbf{Z}_{\mathrm{pp}}|$ and $|\mathbf{Z}_{\mathrm{ss}}|$, so the total sound power output of the control system is approximately,

$$W_{\mathrm{T}} \approx \frac{1}{2}\left[\mathbf{q}_{\mathrm{p}}^{\mathrm{H}}\operatorname{Re}\{\mathbf{Z}_{\mathrm{pp}}\}\mathbf{q}_{\mathrm{p}} + \mathbf{q}_{\mathrm{s}}^{\mathrm{H}}\operatorname{Re}\{\mathbf{Z}_{\mathrm{ss}}\}\mathbf{q}_{\mathrm{s}}\right] \quad (2.40)$$

This indicates that the total sound power output of the control system always increases after the control system is turned on for such a planar virtual sound barrier, where an array of secondary sources is used to control the primary sound field generated by a monopole source at a distance at least a few wavelengths away (Guo, Pan, and Bao, 1997; Guo and Pan, 1997).

There exist some configurations of the control system that can create the largest area of quiet zone with the least increase of total power output. Those configurations correspond to the arrangement where adjacent secondary sources enhance each other at the error sensors, so that the control system requires relatively small sound power to cancel the primary sound field. It is a function of the distances between the primary and secondary sources, secondary source array and error microphone array, and the sound wavelength. A planar virtual sound barrier system with 5×5 equally spaced secondary sources and 5×5 equally spaced error microphones is used as an example for simulations, where the distance between the primary and secondary sources is $5\lambda$, and the separation distance between the secondary sources is the same as that between the error microphones.

Figure 2.4(a) illustrates the total power output increase of the system ($\Delta W_{\mathrm{T}}$) as a function of the normalized secondary source separation distance $d/\lambda$ for different distances between the secondary source plane and error microphone plane of $0.4\lambda$ (solid line), $0.8\lambda$ (dashed line), and $1.6\lambda$ (dotted line). $\Delta W_{\mathrm{T}} = 10\log_{10}(W_{\mathrm{T}}/W_{\mathrm{P}})$, where $W_{\mathrm{T}}$ is the total power output of the system with optimal control and $W_{\mathrm{P}}$ is that without control. It is clear that the minimum total power output increase of the system is a complicated function of many factors, and a good candidate for the secondary source separation distance is a value that is a little bit larger than the half wavelength.

Figure 2.4(b) displays the sound pressure attenuation in the $x$–$z$ plane behind the error microphone array at the frequency corresponding to one wavelength. In the simulation, the separation distance between the secondary sources is $0.4\lambda$ and the distance between the secondary source plane and error microphone plane is also $0.4\lambda$. It can be seen that the sound pressure level behind the error microphone plane is reduced where the quiet zone has the shape of a wedge. Unfortunately, the sound pressure outside the quiet zone is increased, indicating that the total sound power of the system increases after control. As can be observed from the sound distribution pattern in front of the secondary sources, a kind of standing wave is formed

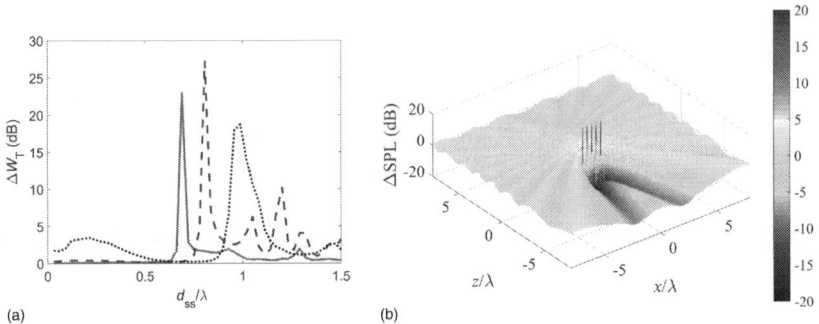

(a)                                                                      (b)

**FIGURE 2.4**
(a) The total power output increase of the control system as a function of the normalized secondary source separation distance, $d_{ss}/\lambda$, for different distances between the secondary source plane and error microphone plane of $0.4\lambda$ (solid line), $0.8\lambda$ (dashed line), and $1.6\lambda$ (dotted line), (b) the sound pressure attenuation in the $x$–$z$ plane of the control system at the frequency corresponding to one wavelength with 5×5 secondary sources and 5×5 error microphones equally spaced in two planes, where the separation distance between the secondary sources and the distance between the secondary source plane and error microphone plane are $0.4\lambda$.

between the forward-going incident primary wave and the backward-going wave generated by the secondary source array, so the mechanism for sound attenuation behind the planar virtual sound barrier by using an array of monopole sources is sound reflection, which reflects the primary incidence sound back in the direction of the incidence wave.

### 2.2.3 Control of General Primary Sound Fields

Practical primary sound fields in a free field might be generated by large machinery in an open space or traffic vehicles along highways, or airplanes in the sky. Some sound sources are limited in size and can be modeled with a group of monopole sources by using the equivalent source method (Johnson et al., 1998), so the principle described in Section 2.2.2 can be applied to investigate the performance of the planar virtual sound barriers in practice.

One difference with practical primary sound fields and those discussed in Sections 2.2.1 and 2.2.2 is that there usually exists a ground. When a nearby reflective surface is introduced, the size of the quiet zone and the total power output of the planar virtual sound barrier system are affected. It has been found that the reflective surface significantly affects the size of the quiet zone and the sound power output of the systems, when compared with the same systems in free space. The effect is related to the heights of the sources and the error sensor above the surface, the distances between the primary and secondary sources, and the distances between the secondary source and the error sensor. While the quiet zones created by most arrangements of the control system in half-space are smaller than those in free space, the quiet zones created by the arrangements of the control system perpendicular to the reflective surface may be larger than those in free space (Guo and Pan, 1998a).

## 2.3 Control of Sound Propagation Through a Finite Size Aperture

There are many practical situations where sound propagates through finite size apertures. For example, doors and windows in buildings, or holes and gaps in enclosures. This section discusses the control of sound propagation through a finite size aperture where there is open space on both sides of the aperture. Section 2.4 discusses the control of sound radiation from the opening of an enclosure, and Section 2.5 investigates the control of sound transmission via an opening into an enclosure.

### 2.3.1 Primary Sound Field with a Finite Size Aperture

Figure 2.5 shows a configuration of the problem where there is a rectangular aperture with dimensions of $2a \times 2b$ located in an infinitely large, rigid wall with a finite thickness of $d$ and a primary point monopole sound source is located outside the aperture.

The incident sound pressure at location $\mathbf{r} = (x, y, z)$ generated by the primary source located at $\mathbf{r}_p = (x_p, y_p, z_p)$ with source strength $q_p$ can be expressed by (Bies, Hansen, and Howard, 2018).

$$p_i(\mathbf{r}) = \frac{jk\rho_0 c_0 q_p}{4\pi |\mathbf{r} - \mathbf{r}_p|} e^{-jk|\mathbf{r} - \mathbf{r}_p|} \tag{2.41}$$

where $k$ is the wavenumber, $\rho_0$ is the density of air, and $c_0$ is the speed of sound. The overall sound pressure at the incident side of the aperture can be calculated by the superposition of the incidence wave, the reflective wave and the scattering wave as (Sgard, Nelisse, and Atalla, 2007)

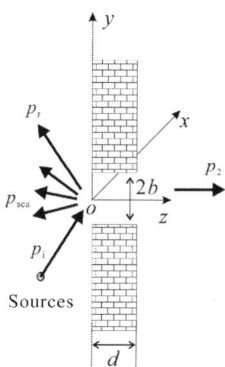

**FIGURE 2.5**
An aperture in an infinitely large, rigid wall with a finite thickness; a primary point monopole sound source is located outside the aperture.

$$p_1(\mathbf{r}) = p_i(\mathbf{r}) + p_r(\mathbf{r}) + p_{sca}(\mathbf{r}) \tag{2.42}$$

The reflective and the scattering waves can be expressed by

$$p_r(\mathbf{r}) = \frac{jk\rho_0 c_0 q_p}{4\pi|\mathbf{r} - \mathbf{r}_r|} e^{-jk|\mathbf{r} - \mathbf{r}_r|} \tag{2.43}$$

$$p_{sca}(\mathbf{r}) = \int_{S_1} G(\mathbf{r}, \mathbf{s}) \frac{1}{j\omega\rho_0} \frac{\partial p_1(\mathbf{s})}{\partial z} dS \tag{2.44}$$

where $S_1$ is the aperture surface on the source side, $\mathbf{r}_r$ is the location of the image source of the primary sound source in relation to the wall, and $G(\mathbf{r}, \mathbf{s})$ is the Green function in the semi-infinitely large space at one side of the wall for a monopole sound source located at $\mathbf{s}$ on the aperture surface,

$$G(\mathbf{r}, \mathbf{s}) = \frac{j\omega\rho_0}{2\pi|\mathbf{r} - \mathbf{s}|} e^{-jk|\mathbf{r} - \mathbf{s}|} \tag{2.45}$$

Similarly, the sound radiated by the aperture into the semi-infinitely large free space behind the infinite wall is given by

$$p_2(\mathbf{r}) = -\int_{S_2} G(\mathbf{r}|\mathbf{s}) \frac{1}{j\omega\rho_0} \frac{\partial p_2(\mathbf{s})}{\partial z} dS \tag{2.46}$$

where $S_2$ is the aperture surface on the radiation side. The sound field inside the aperture (a rigid cavity with two ends open) can be expressed by

$$p_a(\mathbf{r}) = \sum_{p,q} (A_{pq}^+ e^{-jk_{pq}z} + A_{pq}^- e^{jk_{pq}z}) \psi_{pq}(x, y) \tag{2.47}$$

where

$$\psi_{pq}(x, y) = \cos\left[\frac{p\pi(x+a)}{2a}\right]\cos\left[\frac{q\pi(y+b)}{2b}\right] \tag{2.48}$$

$$k_{pq} = \sqrt{k^2 - \left(\frac{p\pi}{2a}\right)^2 - \left(\frac{q\pi}{2b}\right)^2} \tag{2.49}$$

The normal particle velocity on both end surfaces of the aperture is proportional to its pressure gradient in the $z$ direction, which has the form of

$$\left.\frac{\partial p_a(\mathbf{r})}{\partial z}\right|_{z=0} = -\sum_{p,q} jk_{pq}(A_{pq}^+ - A_{pq}^-)\psi_{pq}(x, y) \tag{2.50}$$

$$\left.\frac{\partial p_a(\mathbf{r})}{\partial z}\right|_{z=d} = -\sum_{p,q} jk_{pq}\left(A_{pq}^+ e^{-jk_{pq}d} - A_{pq}^- e^{jk_{pq}zd}\right)\psi_{pq}(x,y) \tag{2.51}$$

By using the normal particle velocity continuity condition on both sides of the aperture at the end surfaces, the sound radiated by the aperture to both sides of the wall can be written as

$$p_{sca}(\mathbf{r}) = -\sum_{p,q}\frac{k_{pq}}{k\rho_0 c_0}\left(A_{pq}^+ - A_{pq}^-\right)\int_0^a\int_0^b G(\mathbf{r}\,|\,(u,v))\psi_{pq}(u,v)dudv \tag{2.52}$$

$$p_2(\mathbf{r}) = \sum_{p,q}\frac{k_{pq}}{k\rho_0 c_0}\left(A_{pq}^+ e^{-jk_{pq}d} - A_{pq}^- e^{jk_{pq}zd}\right)\int_0^a\int_0^b G(\mathbf{r}\,|\,(u,v))\psi_{pq}(u,v)dudv \tag{2.53}$$

By using the pressure continuity condition on both sides of the aperture at the end surfaces, it has

$$p_i(x,y,0) + p_r(x,y,0) + p_{sca}(x,y,0) = \sum_{p,q}\left(A_{pq}^+ + A_{pq}^-\right)\psi_{pq}(x,y) \tag{2.54}$$

$$p_2(x,y,d) = \sum_{p,q}\left(A_{pq}^+ e^{-jk_{pq}d} + A_{pq}^- e^{jk_{pq}d}\right)\psi_{pq}(x,y) \tag{2.55}$$

Substituting the primary incidence pressure $p_i(\mathbf{r})$, the reflective wave $p_r(\mathbf{r})$, and Equations (2.52) and (2.53) into Equations (2.54) and (2.55), multiplying all the equations by $\psi_{mn}(x,y)$, and then integrating over the respective surfaces of the aperture, gives,

$$F_{mn} - \sum_{p,q} jk_{pq}\left(A_{pq}^+ - A_{pq}^-\right)Z_{pqmn} = \left(A_{mn}^+ + A_{mn}^-\right)\Lambda_{mn} \tag{2.56}$$

$$\sum_{p,q} jk_{pq}\left(A_{pq}^+ e^{-jk_{pq}d} - A_{pq}^- e^{jk_{pq}d}\right)Z_{pqmn} = \left(A_{mn}^+ e^{-jk_{mn}d} + A_{mn}^- e^{jk_{mn}d}\right)\Lambda_{mn} \tag{2.57}$$

where $\Lambda_{mn} = \varepsilon_m\varepsilon_n ab$, $\varepsilon_0 = 2$ and $\varepsilon_m = 1$ for $m \neq 0$, and

$$F_{mn} = \int_0^a\int_0^b \frac{jk\rho cq_p}{2\pi|\mathbf{s}-\mathbf{r}_p|}e^{-jk|\mathbf{s}-\mathbf{r}_p|}\psi_{mn}(x,y)dxdy \tag{2.58}$$

$$Z_{pqmn} = \int_{-a}^a\int_{-b}^b\int_{-a}^a\int_{-b}^b \frac{e^{-jk\sqrt{(x-u)^2+(y-v)^2}}}{2\pi\sqrt{(x-u)^2+(y-v)^2}}\psi_{pq}(u,v)\psi_{mn}(x,y)dudvdxdy \tag{2.59}$$

is a specific radiation impedance, which can be calculated numerically or with an approximate expression (Sha, Yang, and Gan, 2005). The modal magnitudes for all aperture modes can be obtained by solving Equations (2.56) and (2.57), which can be further substituted to Equation (2.53) to obtain the transmitted sound. The transmitted power from the opening can be calculated by carrying out an integration of the real part of the sound intensity on the radiation side surface of the aperture as

$$W_a(\mathbf{r}_p) = \frac{1}{2} \mathrm{Re} \left\{ \int_{S_2} p_2(\mathbf{s}) v_2^*(\mathbf{s}) dS \right\} \tag{2.60}$$

where the normal particle velocity in the $z$ direction can be obtained from Equation (2.51).

For a plane wave incident sound, the procedure for the calculation is similar, except the incident sound pressure at location $\mathbf{r}$ is expressed by

$$p_i(\mathbf{r}) = A_p e^{-jk\mathbf{n}\cdot\mathbf{r}} \tag{2.61}$$

where $A_p$ is the amplitude of the primary sound field, $\mathbf{n} = (n_x, n_y, n_z)$ is the unit vector in the propagation direction of the plane wave, and $k$ is the wavenumber.

### 2.3.2 Control of Sound Propagation Through a Finite Size Aperture

Figure 2.6 shows a rectangular aperture with dimensions of $2a \times 2b$ located in an infinitely large, rigid wall with a finite thickness of $d$ where a secondary point monopole sound source for active noise control (ANC) is located inside the aperture.

For a secondary source located at $\mathbf{r}_s = (x_s, y_s, z_s)$ in the aperture shown in Figure 2.6, the sound pressure in the aperture has an extra source term in addition to that in Equation (2.47), and can be expressed as

$$p_s(\mathbf{r}) = \sum_{p,q} \left( A_{pq}^+ e^{-jk_{pq}z} + A_{pq}^- e^{jk_{pq}z} + \frac{\rho \omega q_s e^{-jk_{pq}z - z_s}}{2k_{pq}\Lambda_{pq}} \psi_{pq}(x_s, y_s) \right) \psi_{pq}(x,y) \tag{2.62}$$

Similarly, the modal magnitudes for all aperture modes generated by the secondary source can be calculated as well as the transmitted sound and power to the radiation side of the wall. Because there is no incidence sound from the source side, $p_i(\mathbf{r})$ in Equation (2.42) only contains the scattering term $p_{sca}(\mathbf{r})$ in the calculation.

The total sound pressure with active control is a superposition of both the primary and the control sounds,

$$p_t(\mathbf{r}) = p_p(\mathbf{r}) + Z_{se}(\mathbf{r})q_s \tag{2.63}$$

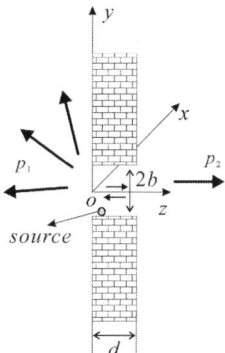

**FIGURE 2.6**
An aperture in an infinitely large, rigid wall with a finite thickness; a secondary point mono-pole sound source is installed inside the aperture.

where $p_p(\mathbf{r})$ is the primary sound pressure at location $\mathbf{r}$, and $Z_{se}(\mathbf{r})$ is the trans-fer function from the source strength of the secondary source to the sound pressure at location $\mathbf{r}$. All these can be calculated with the above established model. The objective of the control is to reduce the total radiation power from the aperture to the radiation side, and different cost functions can be used for obtaining the optimal strength of the secondary source. For exam-ple, using the summation of the squared total sound pressure at $N_e$ error locations as the cost function,

$$J_p = \sum_{i=1}^{N_e} |p_t(\mathbf{r}_i)|^2 + \beta q_s^2 \tag{2.64}$$

where $\beta$ is a positive real number to constrain the strengths of the second-ary source. For multiple secondary sources to reduce the sound pressure at multiple error locations, Equations (2.63) and (2.64) can be written in matrix form as

$$\mathbf{p}_t = \mathbf{p}_p + \mathbf{Z}_{se}\mathbf{q}_s \tag{2.65}$$

$$J_p = \mathbf{p}_t^H \mathbf{p}_t + \beta \mathbf{q}_s^H \mathbf{q}_s \tag{2.66}$$

where the superscript H denotes the Hermitian transpose, $\mathbf{p}_p$ is the vec-tor of the primary sound pressure at $N_e$ error locations, $\mathbf{q}_s$ is the vector of the source strength of $N_s$ secondary sources, while $\mathbf{Z}_{se}$ is the transfer function matrix, whose elements are the transfer functions from all the secondary sources to all the error sensors. The optimal secondary source strength that minimizes the above cost function can be calculated by (Hansen et al., 2013)

$$\mathbf{q}_s^o = -\left(\mathbf{Z}_{se}^H \mathbf{Z}_{se} + \beta \mathbf{I}\right)^{-1} \mathbf{Z}_{se}^H \mathbf{p}_p \qquad (2.67)$$

With the optimal secondary source strength, both the sound field with active control and the total radiation power from the aperture to the radiation side can be calculated. It is also possible to use directly the radiation power from the aperture to the radiation side as the cost function to calculate the optimal secondary source strength, which provides an upper-limit performance for the specified secondary source configuration despite error-sensing strategies and configurations.

### 2.3.3 The Upper-Limit Frequency

Figure 2.7 shows an opening in an infinitely large rigid wall with a finite thickness of 0.1 m. The opening height is $h=0.67$ m, and three different widths ($w=0.01$ m, 0.1 m, and 0.432 m) are considered. Calculations were carried out using the model expansion method introduced in the preceding sections by using 225 modes to ensure the accuracy for frequencies below 4000 Hz. A point monopole primary source is located 20 m from the center of the opening at position F to simulate far-field incidence. Secondary sources $n_h$ and $n_w$ are placed at each of the vertical (height) and horizontal (width) sides of the opening respectively, so the total number of secondary sources is $2(n_h+n_w)$. The sum of the squared sound pressure at 200 evaluation points (adding more evaluation points makes little difference) behind the opening, at a distance of 5 m from the opening center, is used as the cost function to be minimized.

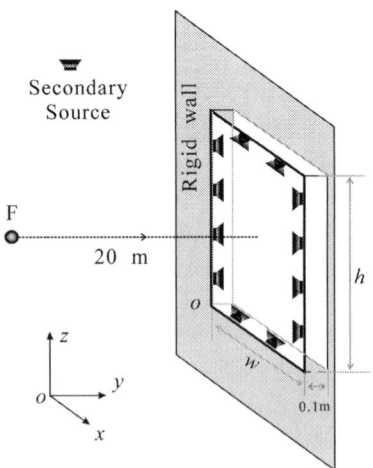

**FIGURE 2.7**
An opening in an infinitely large, rigid wall with a finite thickness of 0.1 m; the opening height is 0.67 m; and the secondary monopole sources are installed in the opening.

Normal incidence is approximated by locating a point monopole primary source at position F. Four cases were investigated, with the secondary sources being installed only on the two opposite sides of the opening: (i) $w=0.01$ m, $n_h=1, 2, 4, 6$, and 10, $n_w=0$; (ii) $w=0.1$ m, $n_h=1, 2, 4, 6$, and 10, $n_w=0$; (iii) $w=0.432$ m, $n_h=1, 2, 4, 6$, and 10, $n_w=0$; (iv) $w=0.432$ m, $n_h=0, n_w=1, 2, 4, 6$, and 10. The secondary source intervals on the vertical and horizontal sides are $d_h=h/(n_h+1)$ and $d_w=w/(n_w+1)$ respectively. The simulation results are shown in Figure 2.8.

In case (i), $w=0.01$ m, $h=0.67$ m, the secondary sources are only installed on the height sides of the opening. The upper-limit effective frequencies (the lowest frequency when noise reduction [NR] is less than 10 dB) for $n_h=1, 2, 4, 6$, and 10 in Figure 2.8(a) are 474 Hz, 1195 Hz, 1598 Hz, 3549 Hz, and 6200 Hz (not shown in the figure), respectively, and the corresponding $kd_h$ are $\pi$, $1.6\pi$, $1.3\pi$, $2\pi$, and $2.2\pi$. It seems that $kd_h$ tends to be $2\pi$ when the number of the secondary sources is sufficiently large for the case where the opening width is much smaller than the secondary source interval. When the width becomes larger, but is still less than $d_h$ (case (ii)),

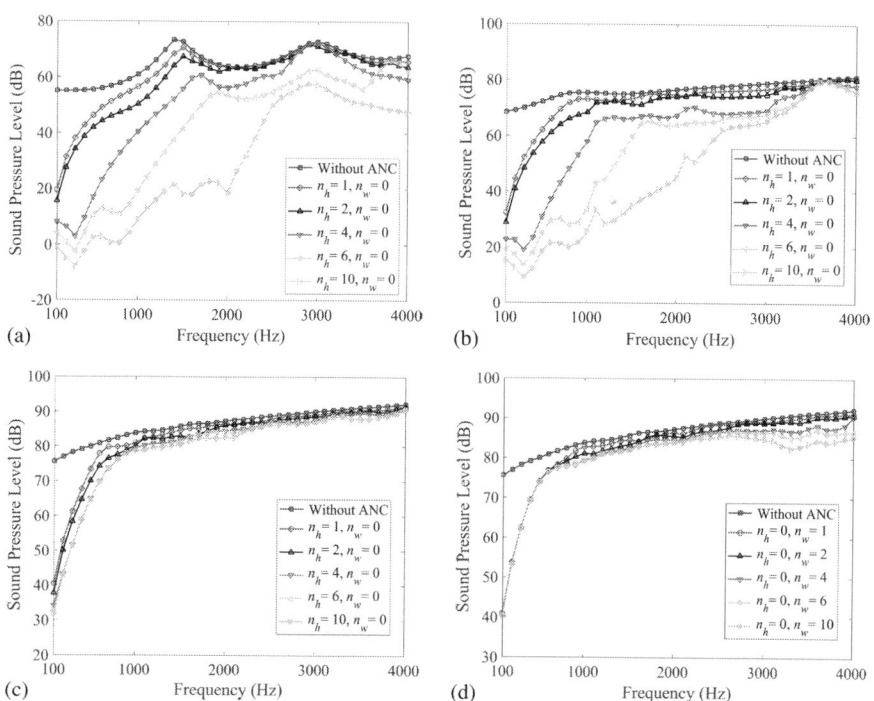

(a) (b) (c) (d)

**FIGURE 2.8**
The averaged, squared sound pressure level without and with control under approximately normal incidence (source point F) for different opening widths and a fixed opening height of 0.67 m: (a) $w=0.01$ m, $n_h=1, 2, 4, 6$, and 10, $n_w=0$; (b) $w=0.1$ m, $n_h=1, 2, 4, 6$, and 10, $n_w=0$; (c) $w=0.432$ m, $n_h=1, 2, 4, 6$, and 10, $n_w=0$; (d) $w=0.432$ m, $n_h=0, n_w=1, 2, 4, 6$, and 10.

the upper-limit frequencies in Figure 2.8(b) for $n_h = 1, 2, 4$, and 6 are 504 Hz, 731 Hz, 1141 Hz, and 1700 Hz, respectively, and the corresponding $kd_h$ are $\pi, 0.9\pi, \pi$, and $\pi$.

When the opening width is larger than the secondary source interval on the vertical sides, the upper-limit frequency depends mainly on the opening width, and its maximum can be estimated by $f_{max} = c_0/w$, i.e., $kw = 2\pi$. For example, the upper-limit frequency for $n_h = 10$ in case (ii) ($d_h = 0.061$ m, $w = 0.1$ m) in Figure 2.8(b) is about 3233 Hz, and the corresponding $kw$ is about $1.9\pi$, while the corresponding $kd_h = 1.2\pi$. In Figure 2.8(c), the upper-limit frequencies are 429 Hz, 499 Hz, 633 Hz, 636 Hz, and 637 Hz for $n_h = 1, 2, 4, 6$, and 10, respectively, and the corresponding $kw$ are in the range of $[1.1\pi, 1.6\pi]$. The upper-limit frequency for $n_h = 40$ (not shown in the figure) is also calculated, which is 670 Hz, corresponding to $1.7\pi$. It seems that the value tends to be $2\pi$ when the number of secondary sources is sufficiently large. The same trend can be found in Figure 2.8(d) where the height is larger than the secondary source interval on the wide sides, the $kh$ is about $1.6\pi$ for these five values of $n_w$ and is $1.8\pi$ for $n_w = 40$ (not shown in the figure).

A surface control system has the secondary sources distributed evenly over the entire opening, while a boundary control system only installs the secondary sources on its frame and thus has advantages of better ventilation and easier access. When sufficient secondary sources are placed on all sides of the opening boundaries, the upper-limit frequency of the boundary control systems is mainly decided by the length of the narrow side of the opening. Figure 2.9(a) shows that the upper-limit frequency for a narrow opening ($w = 0.1$ m, $h = 0.67$ m) is about 3500 Hz, corresponding to $kw = 2\pi$. If the number of the secondary sources is limited, putting more secondary sources on the long sides can achieve higher upper-limit frequencies than putting the secondary sources on the short sides of the opening. For example, if only 16 secondary sources can be used, Figure 2.9(a) shows that

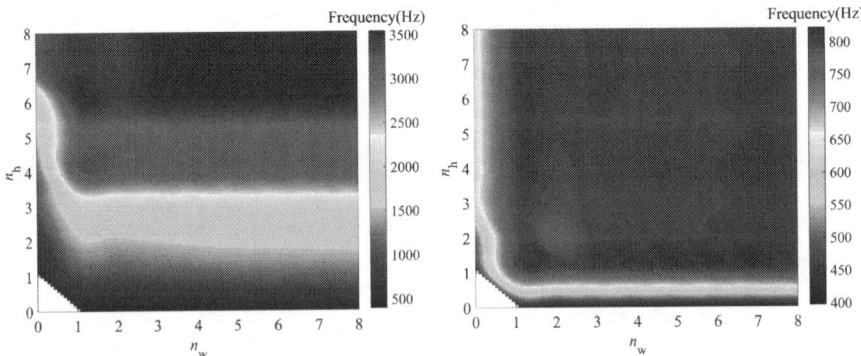

**FIGURE 2.9**
The contour map of the upper-limit frequencies of the virtual sound barrier system for $n_h = 0 \sim 8, n_w = 0 \sim 8$ (excluding $n_h = n_w = 0$) with different opening size: (a) $w = 0.1$ m, $h = 0.67$ m, (b) $w = 0.432$ m, $h = 0.67$ m.

the upper-limit frequency can achieve 3347 Hz with $n_h=6$ and $n_w=2$, but only 1630 Hz with $n_h=2$ and $n_w=6$. When the width of an opening is similar to the height, the same trend has been found, as shown in Figure 2.9(b) where the upper-limit frequency is about 798 Hz for $w=0.432$ m, $h=0.67$ m, corresponding to $kw=2\pi$.

Table 2.1 further compares the upper-limit frequencies of the virtual sound barrier systems with 16 secondary sources distributed evenly over the entire opening (surface control) or on the boundaries for two openings with different sizes. It is clear that the upper-limit frequency of the boundary control systems with the secondary sources on all sides is higher than those with the secondary sources on two opposite sides only. To reach the maximal upper-limit frequency corresponding to $kw=2\pi$, the number of the sources on the long side should be sufficient. The upper-limit frequency of the surface control system is higher than that, with the boundary control systems for a normal opening, such as $w=0.432$ m. But for a narrow opening, such as $w=0.1$ m, the upper-limit frequency is lower than that of the boundary control system provided that sufficient sources are installed on the long sides.

Oblique incidence and grazing incidence can be considered similarly (Wang, Tao, and Qiu, 2019). The surface control system can control sound propagation through a finite aperture up to any frequency as long as sufficient secondary sources are used and their separation distances are smaller than the half wavelength of the sound with the highest frequency.

**TABLE 2.1**

The Upper-Limit Frequencies of the Planar Virtual Sound Barrier Systems with 16 Secondary Sources Distributed Evenly on the Entire Opening or on the Boundaries

| Configurations | The Upper-limit Frequency (Hz) for Openings | |
| --- | --- | --- |
| | 0.1 m wide, 0.67 m high | 0.432 m wide, 0.67 m high |
| Boundary control ($n_h=0$, $n_w=8$) | 511 | 400 |
| Boundary control ($n_h=1$, $n_w=7$) | 1178 | 787 |
| Boundary control ($n_h=2$, $n_w=6$) | 1630 | 796 |
| Boundary control ($n_h=4$, $n_w=4$) | 3080 | 793 |
| Boundary control ($n_h=6$, $n_w=2$) | 3347 | 791 |
| Boundary control ($n_h=7$, $n_w=1$) | **3381** | 792 |
| Boundary control ($n_h=8$, $n_w=0$) | 3148 | 636 |
| Surface control (even distribution on the opening) | 2688 | **1208** |

The boundary control system can effectively control sound transmission through the opening up to a certain frequency. This upper-limit effective frequency is decided by the length of the short sides of the opening, which corresponds to the wavelength of the sound that can be controlled.

If the secondary sources of the boundary control systems are only placed on the long sides of a narrow opening, the upper-limit frequency can be estimated by $f_{max} = c_0/d_h$ when the width of the opening is much smaller than the secondary source interval $d_h$. The upper-limit frequency reduces to $f_{max} = c_0/2d_h$ when the width becomes larger but still less than $d_h$. When the width $w$ is larger than the secondary source interval, the maximal upper-limit frequency can be estimated by $f_{max} = c_0/w$ provided that plenty of the secondary sources are used. Placing secondary sources on all sides of the opening is preferable for higher upper-limit frequency, and the maximal upper-limit frequency is decided by $w$ with $f_{max} = c_0/w$ where $w$ is the length of the short sides of the opening. For an opening with its width similar to the height, the upper-limit frequency under the grazing incidence is larger than that under normal incidence.

The performance of a boundary control system can be better than that of a surface control system when the number of secondary sources is limited or when the opening is narrow. The performance of a boundary control system does not vary significantly for oblique incidence with large incident angles.

## 2.4 Control of Sound Radiation from an Opening of an Enclosure

In many practical applications, sound sources in an enclosure radiate noise outside via the openings of the enclosure. An analytical model for calculating the sound field inside and outside of the enclosure is introduced in this section, and then secondary sources are placed on the openings to construct a planar virtual sound barrier to reduce the sound radiation from the opening (Wang, Tao, and Qiu, 2015).

### 2.4.1 Sound Radiation from an Opening of an Enclosure

Figure 2.10 shows a schematic diagram of a planar virtual sound barrier system implemented in an open cavity with dimensions of $l_x$ long, $l_y$ wide, and $l_z$ high. The origin of the coordinate is at the back left bottom vertex of the cavity. All the walls of the cavity are rigid except the opening. The planar virtual sound barrier system consists of arrays of loudspeakers and microphones, which are distributed evenly on two different planes parallel to the opening surface.

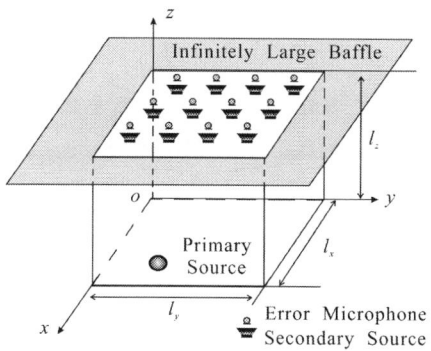

**FIGURE 2.10**
A planar virtual sound barrier system with an array of secondary sources on the baffled opening of a rigid cavity.

The sound pressure at location **r** (omitting the time dependence $e^{j\omega t}$) excited by a point monopole source inside the cavity can be obtained by solving the inhomogeneous Helmholtz equation (Morse and Ingard, 1968)

$$(\nabla^2 + k^2)p = -j\rho_0\omega q_0\delta(\mathbf{r} - \mathbf{r}_s) \tag{2.68}$$

where $k$ is the wavenumber, $\omega$ is the angular frequency, $q_0$ is the volume velocity of the source, and $\mathbf{r}_s$ is the position of the monopole source. The solution of Equation (2.68) can be written as (Huang, Qiu, and Kang, 2011)

$$p_{in}(\mathbf{r}, \omega) = \sum_{n=1}^{N} (P_n^+ e^{-jk_{nz}z} + P_n^- e^{jk_{nz}z})\phi_n(x, y) + \iiint_V j\rho_0\omega q(\mathbf{r}_s)G_A dV \tag{2.69}$$

where $P_n^+$ and $P_n^-$ correspond to the coefficients of the $n$th mode of sound propagating in the positive and negative $z$ directions. The wavenumber $k_{nz}$ and $k$ are related by

$$k_{nz}^2 + \left(\frac{n_x\pi}{l_x}\right)^2 + \left(\frac{n_y\pi}{l_y}\right)^2 = k^2 \tag{2.70}$$

$\phi_n(x, y)$ is the eigenfunction of the $(n_x, n_y)$ mode of an infinitely long, rectangular, rigid duct, which is given by

$$\phi_n(x, y) = \cos\frac{n_x\pi}{l_x}x\cos\frac{n_y\pi}{l_y}y \tag{2.71}$$

and the corresponding modal frequency is

$$f_n = \frac{c_0}{2}\sqrt{\left(\frac{n_x}{l_x}\right)^2 + \left(\frac{n_y}{l_y}\right)^2} \tag{2.72}$$

where $c_0$ is the sound speed.

The Green's function in the duct with the two ends open can be described by (Morse and Ingard, 1968)

$$G_A(\mathbf{r}, \mathbf{r}_0) = \frac{-j}{2S} \sum_{n=1}^{N} \frac{\phi_n(x, y)\phi_n(x_0, y_0)}{\Lambda_n k_{nz}} e^{-jk_{nz}|z-z_0|} \tag{2.73}$$

where $\mathbf{r}_0 = (x_0, y_0, z_0)$ is a point source location in the duct, $S = l_x l_y$ is the area of the opening, and $\Lambda_n$ is defined by

$$\Lambda_n = \begin{cases} 1, & n_x = n_y = 0 \\ \dfrac{1}{2}, & n_x = 0, n_y \neq 0, \text{ or } n_x \neq 0, n_y = 0 \\ \dfrac{1}{4}, & n_x \neq 0, n_y \neq 0 \end{cases} \tag{2.74}$$

The sound pressure outside the cavity, $p_{out}$, can be calculated using the Rayleigh integral when the opening is placed in an infinitely large rigid baffle (Pan and Bies, 1990)

$$p_{out}(\mathbf{r}, \omega) = \frac{1}{2\pi} \iint_{S_1} \frac{\partial p}{\partial \mathbf{n}_{out}}\bigg|_{z=l_z} \left(\frac{e^{-jk \cdot r}}{r}\right) ds_0 \tag{2.75}$$

where $S_1$ stands for the opening surface and $r$ is the distance between $\mathbf{r}$ and a source location, $\mathbf{r}_0 = (x_0, y_0, l_z)$, on the opening surface, which can be calculated by

$$r = \sqrt{(x - x_0)^2 + (y - y_0)^2 + (z - l_z)^2} \tag{2.76}$$

The boundary conditions at $z = 0$ and $z = l_z$ are

$$\frac{\partial p_{in}(\mathbf{r})}{\partial \mathbf{n}_{in}}\bigg|_{z=0} = 0 \tag{2.77}$$

$$p_{in}(\mathbf{r})\big|_{z=l_z} = p_{out}(\mathbf{r})\big|_{z=l_z} \tag{2.78}$$

$$\frac{\partial p_{in}(\mathbf{r})}{\partial \mathbf{n}_{in}}\bigg|_{z=l_z} = \frac{\partial p_{out}(\mathbf{r})}{\partial \mathbf{n}_{out}}\bigg|_{z=l_z} \tag{2.79}$$

Combining Equation (2.69) and Equation (2.77) leads to

$$jk_{mz}\left(-P_m^+ + P_m^-\right) + \rho_0 \omega q_0 \frac{\phi_m(x_s, y_s)}{2\Lambda_m S} je^{-jk_{mz}z_s} = 0 \tag{2.80}$$

The continuity of sound pressure and particle velocity at the opening leads to the mode-matching equations between the interior pressure field and exterior pressure field

$$\sum_{n=1}^{N}\left[k_{nz}\left(-P_n^+e^{-jk_{nz}l_z}+P_n^-e^{jk_{nz}l_z}\right)-\rho_0\omega q_0\frac{\phi_n(x_s,y_s)}{2\Lambda_nS}e^{-jk_{nz}(l_z-z_s)}\right]\frac{S}{\rho_0\omega}Z_{nm}$$

$$=-\left[P_m^+e^{-jk_{mz}l_z}+P_m^-e^{jk_{mz}l_z}+\rho_0\omega q_0\frac{\phi_m(x_s,y_s)}{2\Lambda_mSk_{mz}}e^{-jk_{mz}(l_z-z_s)}\right]\Lambda_mS \tag{2.81}$$

where

$$Z_{nm}=\frac{j\rho_0\omega}{2\pi S}\int_0^{l_x}\int_0^{l_y}\int_0^{l_x}\int_0^{l_y}\phi_n(x_1,y_1)\left(\frac{e^{-jk\cdot\sqrt{(x_0-x_1)^2+(y_0-y_1)^2}}}{\sqrt{(x_0-x_1)^2+(y_0-y_1)^2}}\right)\phi_m(x_0,y_0)dx_1dy_1dx_0dy_0 \tag{2.82}$$

is the normalized radiation impedance (Mangiarotty, 1963; Burnett and Soroka, 1972).

Using Equations (2.80) and (2.81), vectors $P_N^+=[P_1^+,P_2^+,\ldots,P_N^+]^T$ and $P_N^-=[P_1^-,P_2^-,\cdots,P_N^-]^T$ can be represented by

$$P_N^-=(Z_{A1}+Z_{A2}+Z_{P1}+Z_{P2})^{-1}(-Z_{P1}C_{N3}-Z_{A1}C_{N3}-C_{N1}-C_{N2}) \tag{2.83}$$

$$P_N^+=(Z_{A1}+Z_{A2}+Z_{P1}+Z_{P2})^{-1}$$
$$(-Z_{P1}C_{N3}-Z_{A1}C_{N3}-C_{N1}-C_{N2})+C_{N3} \tag{2.84}$$

respectively, with

$$Z_{A1}=\frac{-S}{\rho_0\omega}\begin{pmatrix}k_{1z}e^{-jk_{1z}l_z}Z_{11} & k_{2z}e^{-jk_{2z}l_z}Z_{21} & \cdots & k_{Nz}e^{-jk_{Nz}l_z}Z_{N1}\\ k_{1z}e^{-jk_{1z}l_z}Z_{12} & k_{2z}e^{-jk_{2z}l_z}Z_{22} & \cdots & k_{Nz}e^{-jk_{Nz}l_z}Z_{N2}\\ \vdots & \vdots & \ddots & \vdots\\ k_{1z}e^{-jk_{1z}l_z}Z_{1N} & k_{2z}e^{-jk_{2z}l_z}Z_{2N} & \cdots & k_{Nz}e^{-jk_{Nz}l_z}Z_{NN}\end{pmatrix} \tag{2.85}$$

$$Z_{A2}=\frac{S}{\rho_0\omega}\begin{pmatrix}k_{1z}e^{jk_{1z}l_z}Z_{11} & k_{2z}e^{jk_{2z}l_z}Z_{21} & \cdots & k_{Nz}e^{jk_{Nz}l_z}Z_{N1}\\ k_{1z}e^{jk_{1z}l_z}Z_{12} & k_{2z}e^{jk_{2z}l_z}Z_{22} & \cdots & k_{Nz}e^{jk_{Nz}l_z}Z_{N2}\\ \vdots & \vdots & \ddots & \vdots\\ k_{1z}e^{jk_{1z}l_z}Z_{1N} & k_{2z}e^{jk_{2z}l_z}Z_{2N} & \cdots & k_{Nz}e^{jk_{Nz}l_z}Z_{NN}\end{pmatrix} \tag{2.86}$$

$$Z_{P1}=S\begin{pmatrix}\Lambda_1e^{-jk_{1z}l_z} & 0 & & \\ 0 & \Lambda_2e^{-jk_{2z}l_z} & & \\ & & \ddots & \\ & & & \Lambda_Ne^{-jk_{Nz}l_z}\end{pmatrix} \tag{2.87}$$

$$Z_{P2} = S \begin{pmatrix} \Lambda_1 e^{jk_{1z}l_z} & 0 & & \\ 0 & \Lambda_2 e^{jk_{2z}l_z} & & \\ & & \ddots & \\ & & & \Lambda_N e^{jk_{Nz}l_z} \end{pmatrix} \tag{2.88}$$

$$C_{N1} = \frac{-q_0}{2} \begin{pmatrix} Z_{11} & Z_{21} & \cdots & Z_{N1} \\ Z_{12} & Z_{22} & \cdots & Z_{N2} \\ \vdots & \vdots & \ddots & \vdots \\ Z_{1N} & Z_{2N} & \cdots & Z_{NN} \end{pmatrix} \begin{pmatrix} \dfrac{\phi_1(x_s,y_s)}{\Lambda_1} e^{-jk_{1z}(l_z-z_s)} \\ \dfrac{\phi_2(x_s,y_s)}{\Lambda_2} e^{-jk_{2z}(l_z-z_s)} \\ \vdots \\ \dfrac{\phi_N(x_s,y_s)}{\Lambda_N} e^{-jk_{Nz}(l_z-z_s)} \end{pmatrix} \tag{2.89}$$

$$C_{N2} = \frac{\rho_0\omega q_0}{2} \begin{pmatrix} \dfrac{\phi_1(x_s,y_s)}{k_{1z}} e^{-jk_{1z}(l_z-z_s)} \\ \dfrac{\phi_2(x_s,y_s)}{k_{2z}} e^{-jk_{2z}(l_z-z_s)} \\ \vdots \\ \dfrac{\phi_N(x_s,y_s)}{k_{Nz}} e^{-jk_{Nz}(l_z-z_s)} \end{pmatrix} \tag{2.90}$$

$$C_{N3} = \frac{\rho_0\omega q_0}{2S} \begin{pmatrix} \dfrac{\phi_1(x_s,y_s)}{\Lambda_1 k_{1z}} e^{-jk_{1z}z_s} \\ \dfrac{\phi_2(x_s,y_s)}{\Lambda_2 k_{2z}} e^{-jk_{2z}z_s} \\ \vdots \\ \dfrac{\phi_N(x_s,y_s)}{\Lambda_N k_{Nz}} e^{-jk_{Nz}z_s} \end{pmatrix} \tag{2.91}$$

Hence, the sound pressure inside and outside the cavity can be calculated by using Equations (2.69) and (2.75).

### 2.4.2 Surface Control

With the analytical model based on the modal superposition method proposed above, the sound field inside and outside a rectangular cavity with an opening can be calculated. The primary sound pressure vector at $L$ error points can be calculated by

$$\mathbf{p}_\mathrm{p} = \mathbf{Z}_{pe}q_\mathrm{p} \tag{2.92}$$

where $\mathbf{Z}_{pe}$ is the vector of the acoustic transfer functions from the primary source to the $L$ error points, and $q_\mathrm{p}$ is the primary source strength. The sound pressure vector generated by $M$ secondary sources of the planar virtual sound barrier can be calculated by

$$\mathbf{p}_\mathrm{s} = \mathbf{Z}_{se}\mathbf{q}_\mathrm{s} \tag{2.93}$$

where $\mathbf{Z}_{se}$ is the $L \times M$ matrix of acoustic transfer functions from the $M$ secondary sources to the $L$ error points, and $\mathbf{q}_\mathrm{s} = [q_1, q_2, ..., q_M]^\mathrm{T}$ is the strength vector of the secondary sources. The total sound pressure vector for the planar virtual sound barrier system at the error points is

$$\mathbf{p}_\mathrm{t} = \mathbf{p}_\mathrm{p} + \mathbf{p}_\mathrm{s} \tag{2.94}$$

The optimal secondary source strength can be calculated by minimizing the following cost function, which is defined as the sum of the total squared sound pressure at the error points plus the weighted secondary source power

$$J = \mathbf{p}_\mathrm{t}^\mathrm{H}\mathbf{p}_\mathrm{t} + \beta\mathbf{q}_\mathrm{s}^\mathrm{H}\mathbf{q}_\mathrm{s} \tag{2.95}$$

where $\beta$ is a positive real number for constraining the control effort. A small value of $\beta$ is helpful to increase the noise reduction of the system but the strengths of the secondary sources may be too large for practical application. A rule of thumb to select $\beta$ is to set it as $1/5000$ to $1/1000$ of the largest eigenvalue of the matrix $\mathbf{Z}_{se}^\mathrm{H}\mathbf{Z}_{se}$. The optimized strengths of the secondary sources are (Hansen et al., 2013)

$$\mathbf{q}_\mathrm{s} = -(\mathbf{Z}_{se}^\mathrm{H}\mathbf{Z}_{se} + \beta\mathbf{I})^{-1}\mathbf{Z}_{se}^\mathrm{H}\mathbf{Z}_{qe}q_\mathrm{p} \tag{2.96}$$

where $\mathbf{Z}_{pe}$ and $\mathbf{Z}_{se}$ can be obtained by theoretical calculations, numerical simulations, and/or experimental measurements. The analytical model introduced in Section 2.4.1 is a theoretical method that can be used to calculate these transfer functions based on the modal superposition method. The acoustic transfer functions $\mathbf{Z}_{pe}$ and $\mathbf{Z}_{se}$ are calculated with the model first, then the optimal strengths of the secondary and the noise reduction of the planar virtual sound barrier system can be obtained.

The performance of the planar virtual sound barrier system is defined as the noise reduction at $N_\mathrm{v}$ evaluation points outside the opening

$$\mathrm{NR} = 10\log_{10} \frac{\displaystyle\sum_{i=1}^{N_v}|p_p(\mathbf{r}_{v,i})|^2}{\displaystyle\sum_{i=1}^{N_v}|p_t(\mathbf{r}_{v,i})|^2} \qquad (2.97)$$

where $p_p(\mathbf{r}_{v,i})$ and $p_t(\mathbf{r}_{v,i})$ are the sound pressure vectors at the evaluation points in the far field without and with the control.

Figure 2.11 illustrates a virtual sound barrier system consisting of 6 secondary sources and 12 error microphones installed at the opening of a cavity with dimensions 0.432 m long, 0.67 m wide, and 0.598 m high. The interval of secondary sources is 0.232 m in the $x$ direction and 0.235 m in the $y$ direction, and the plane of the secondary sources is at a height of $z=0.498$ m, 0.1 m lower than the opening surface. The plane of the 12 error microphones (error points) is on the opening. A total of 10 points distributed on a semi-sphere with a radius of 1.5 m are used by Equation (2.97) as the evaluation points to estimate the noise power reduction of the system (ISO 3744: 1994). In the simulations, a monopole primary source generates a pure tonal signal of a single frequency from 1 Hz to 500 Hz with a step size of 1 Hz.

The sound power levels with and without control (dotted line and dashed line) and the noise reduction (solid line) of the virtual sound barrier system with $\beta=0$ are shown in Figure 2.12. It can be observed that the system usually has high noise reductions at the modal frequencies of the open cavity, while there is little noise reduction at certain frequencies when the primary

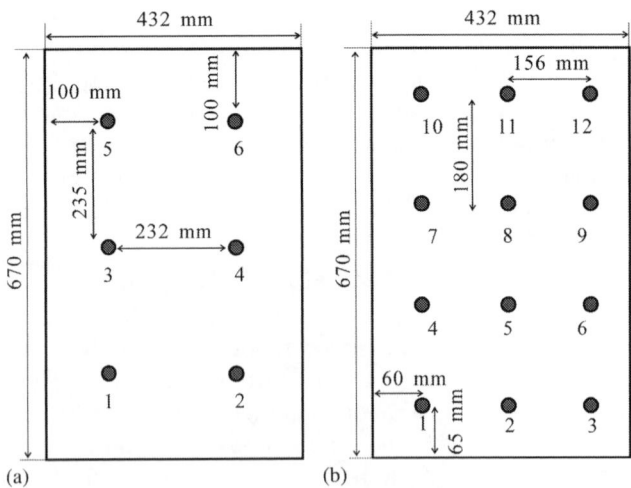

**FIGURE 2.11**
The physical configuration of the virtual sound barrier systems: (a) distribution of 6 secondary sources on the $z=0.498$ m plane, (b) distribution of 12 error microphones on the $z=0.598$ m plane.

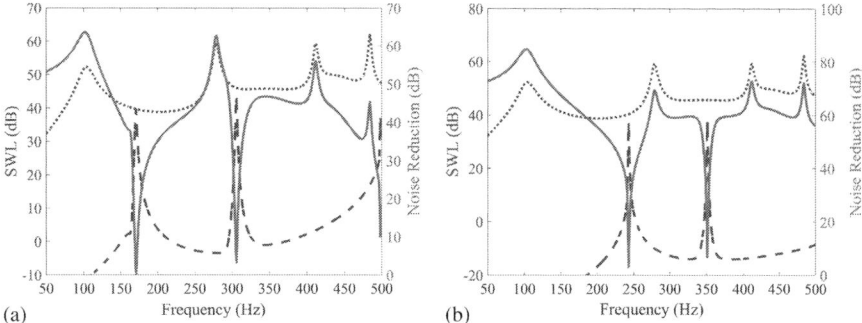

**FIGURE 2.12**
The sound power level with and without control (dotted line and dashed line), and the noise reduction (solid line) with the secondary sources on: (a) $z = 0.498$ m plane, (b) $z = 0.35$ m plane.

sound power level is low or when it is hard for the secondary sources to generate sound. For example, the system has almost zero attenuation at 171 Hz in Figure 2.12(a) when the secondary sources are on the plane of $z = 0.498$ m because they cannot radiate effectively at 171 Hz. A quarter of the wavelength of 171 Hz is about 0.498 m, which is exactly the distance from the plane of secondary sources to the bottom of the open cavity. The sound propagates in the negative $z$ direction and is reflected back to the plane of the secondary source with the opposite phase; as a result, the sound radiation outward at 171 Hz is much less compared to those at other frequencies. If the secondary sources are on a different plane, more noise reduction at 171 Hz can be achieved, as evidenced in Figure 2.12(b), with the 6 secondary sources moved to the plane of $z = 0.35$ m. However, the noise reduction at 243 Hz is only 3.4 dB now for the same reason, that 0.35 m is about a quarter of the wavelength of 243 Hz.

Both the simulation and the experiment results show that the planar virtual sound barriers can effectively reduce sound transmission through the opening except at certain special frequencies that are related to the position of the secondary source plane, which has low sound radiation capability. Three mechanisms have been found to act together to reduce the sound radiation through the opening, which include changing the impedance of the primary source, modal control, and modal rearrangement (Wang, Tao, and Qiu, 2015).

### 2.4.3 Boundary Control

For the surface control system discussed in Section 2.4.2, sound radiation through the opening can be reduced completely by the planar active sound barrier provided that there are sufficient secondary loudspeakers distributed over the entire opening. In some applications, the secondary source loudspeakers are required to be installed only at the edge of the opening to facilitate better access, light, and air circulation through the opening. Because single-layered loudspeakers can only achieve global control of sound radiation through openings at relatively low frequencies, more layers of loudspeakers are used to effectively block sound transmission through openings over a wider frequency band (Wang et al., 2017; Wang, Tao, and Qiu, 2017).

Figure 2.13(a) shows a schematic diagram of a boundary control system, which has a double-layered loudspeaker array fixed at the edge of the opening in the virtual sound barrier system, where the projections of the loudspeakers in both layers on the ground are the same. A rigid rectangular cavity of 0.432 m × 0.670 m × 0.598 m (length × width × height) is used as the model, and the opening size is 0.432 m × 0.670 m. The origin of the coordinate is at the left rear vertex at the bottom of the cavity. In the system, 16 out of the 32 loudspeakers are at the height of 0.448 m while the other 16 are at the height of 0.548 m. Their positions in the $x$–$y$ plane are shown in Figure 2.13(b). The primary sound source that generates the unwanted sound (to be blocked by the proposed system) is assumed to be at (0.1, 0.1, 0.1) m inside the open cavity.

The sound power levels (SWLs) radiated outward through the opening without (dotted line) and with the boundary control system (solid line), are shown in Figure 2.14(a), where the simulation results of the surface control system (dashed line) with 32 loudspeakers evenly distributed over the entire opening are also presented for comparison. The sound power reductions from 450 Hz to 1000 Hz are all more than 40 dB with the boundary control system. There exist some frequencies at which the sound cannot be attenuated well by the surface control system, but the boundary control system does not have the same weakness. For example, the sound reduction at 570 Hz is only 20.7 dB with the surface control system, while it is 54.4 dB at the same frequency with the boundary control system. The boundary control

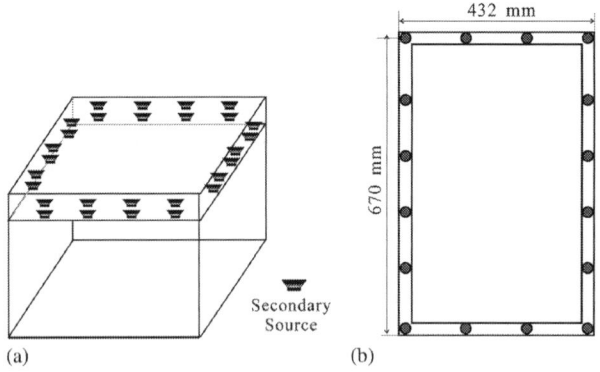

**FIGURE 2.13**

(a) A virtual sound barrier system with double-layer secondary sources on the boundaries of the opening, (b) positions of 16 out of the 32 secondary sources on the $x$–$y$ plane in the virtual sound barrier system.

system can achieve more consistent sound reduction performance than the surface control system over a wider frequency band.

The measured sound power levels in an anechoic chamber without (dotted line and dash-dotted line) and with the virtual sound barrier systems are shown in Figure 2.14(b). Due to the background noise and the dynamic range of the active controller, the noise reduction in the experiments is much less than that in the simulations, but the trends of the curves in Figure 2.14(b) are similar to the simulation curves in Figure 2.14(a), and they lead to similar conclusions as those in the simulations. The surface control system (dashed line) achieves good sound reduction performance at most frequencies, but at some frequencies, such as 510 Hz, the system is hard to converge, which results in a much lower sound power reduction of only about 5 dB. The boundary control system (solid line) achieved a more consistent sound power reduction of more than 15 dB at most frequencies below 1000 Hz. The system is also effective when the primary sound field is in general and more complicated forms because a complicated sound source can be decomposed into the summation of a number of point monopole sources (Wang, Tao, and Qiu, 2017).

The calculation of the sound power of the primary sound source and secondary sources without and with control shows that the sound power of the primary sound source is significantly reduced when the virtual sound barrier is working, which indicates that the main mechanism of active control is unloading the primary source. Figure 2.15 shows the sound pressure level without and with the virtual sound barrier in a $y$–$z$ or $x$–$z$ plane inside and outside the open cavity. It is clear that the sound pressure level outside the open cavity (proportional to the total sound power of the system) is significantly reduced, while inside the cavity the sound pressure level remains the same or even increases.

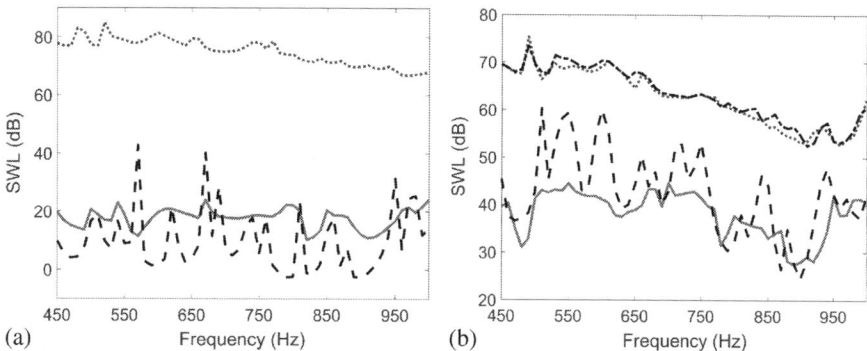

**FIGURE 2.14**
The SWL without (dotted line, and dash-dotted line only for the experiment) and with the surface control (dashed line) and the boundary control (solid line) of the virtual sound barrier system: (a) simulation results, (b) experimental results.

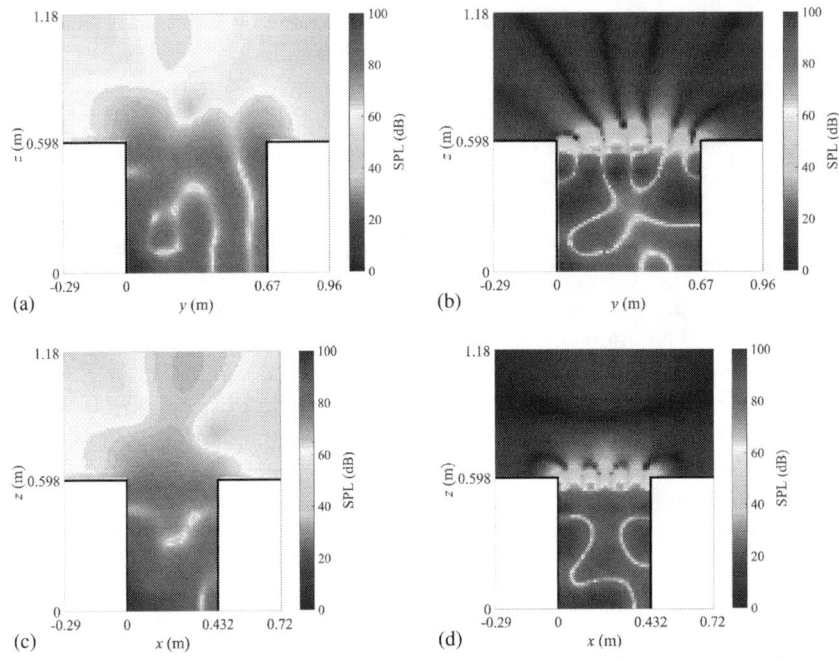

**FIGURE 2.15**
The SPL on the $y$–$z$ and $x$–$z$ planes: (a) without the virtual sound barrier, $x=0.03$ m plane, (b) with the virtual sound barrier, $x=0.03$ m plane, (c) without the virtual sound barrier, $y=0.03$ m plane, (d) with the virtual sound barrier, $y=0.03$ m plane.

### 2.4.4 The Upper-Limit Frequency

The double-layer edge system (DLES) achieves better noise reduction performance than the single-layer edge system (SLES) because the secondary sources at the edge of the same layer cannot excite some modes effectively while those at the other height in the DLES can compensate for it (Wang et al., 2018). The secondary sources in the middle of the opening in the surface control system can excite more modes than those at the edge, so can achieve higher noise reduction over a wider frequency range. There exists an upper-limit frequency for the SLES, DLES, and 3LES (triple-layer edge system). Using more secondary source layers increases the upper-limit frequency of the system.

In the simulations, the dimension of the open cavity is 0.4 m×1.0 m×1.5 m ($l_x \times l_y \times l_z$), and the size of the opening is 0.4 m×1.0 m. Using 20 dB as the threshold, the highest frequency at which the sound power reduction is more than 20 dB is defined as $f_{20}$. Figure 2.16 shows the half wavelength of the sound at $f_{20}$ as a function of the interval between secondary sources in the surface control system, for the SLES, DLES, and 3LES. In the surface control system, the secondary sources are evenly distributed at $z=1.4$ m with the

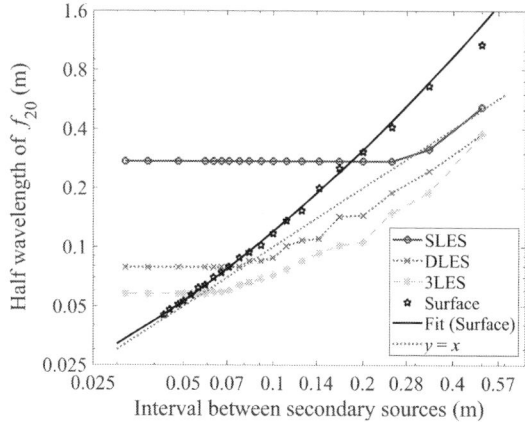

**FIGURE 2.16**
The half wavelength of the sound at $f_{20}$ as a function of the interval between the second-ary sources in the systems with surface control ("Surface" in the figure legend), SLES, DLES and 3LES.

same intervals along both $x$ and $y$ axes. In the SLES, DLES, and 3LES, the sec-ondary sources are evenly distributed along the boundaries. It can be seen that $f_{20}$ of the surface control system increases when the interval between secondary sources becomes smaller and sound radiation at any frequency can be reduced as long as there are sufficient secondary sources. The solid line is the fitted curve for the surface control system and the half wavelength of $f_{20}$ is close to the interval between the secondary sources, especially when the interval between secondary sources is small.

Unlike the surface control system, the half wavelength of the sound at $f_{20}$ is flat for the boundary control systems when the interval between the sec-ondary sources is small, and starts to increase when the interval is about 0.2 m (one or half the short side length of the opening). A small wavelength means a high frequency. The flat half wavelength of the sound at $f_{20}$ with the small secondary source intervals indicates that there exists an upper-limit frequency that is the highest frequency at which the sound radiation can be effectively reduced with sufficient secondary sources.

Figure 2.17(a) shows the upper-limit frequency of the SLES, DLES, and 3LES as a function of $l_x$ when $l_y = 1.0$ m and $l_z = 1.5$ m. It is clear that more lay-ers increase the upper-limit frequency, and that the upper-limit frequency increases significantly when $l_x$ is much smaller than $l_y$, provided that there are sufficient secondary sources at the edge to achieve the best noise reduc-tion performance. The upper-limit frequency depends mainly on the short side of the opening because it changes little after $l_x$ increases to 1.0 m. This is further demonstrated by the results for a narrow opening ($l_x$ is much smaller than $l_y$) shown in Figure 2.17(b), where the sound wavelengths correspond-ing to the upper-limit frequency of the SLES and DLES are about $l_x$ and $l_x/2$,

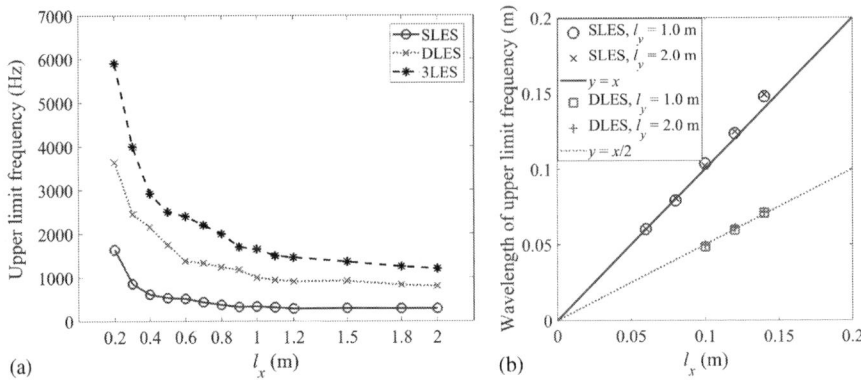

**FIGURE 2.17**
(a) The upper-limit frequency of the virtual sound barrier systems with SLES, DLES, and 3LES as a function of $l_x$ when $l_y = 1.0$ m and $l_z = 1.5$ m, (b) the wavelength of the sound at the upper-limit frequency as a function of $l_x$ when $l_y = 1.0$ m and $2.0$ m, and $l_z = 1.5$ m.

respectively, and remain the same with different $l_y$. Therefore, the upper-limit frequency mainly depends on the length of the short side of the opening, especially for a narrow opening.

## 2.5 Control of Sound Transmission via an Opening into an Enclosure

Many methods have been proposed to reduce sound transmission via an opening into an enclosure. The plenum windows, which are partially opened double-glazed windows with the two openings on opposite sides of the window, were tested in the laboratory in 1973, when it was found that the noise insulation performance of plenum windows with a certain opening size could match that of a closed single glazing (Ford and Kerry, 1973). About 40 years later, Tong et al. (2015) measured the insertion loss of a scale model in the presence of a line source, and then they carried out a full-scale field study along a busy trunk road and found that the noise insulation could be increased by up to 9.5 dBA by using plenum windows. Sound absorption materials such as transparent micro-perforated absorbers were used inside the plenum windows to increase noise attenuation, but the performance improvement in the low-frequency range was limited (Kang and Brocklesby, 2005). The plenum openings can be combined with ANC techniques to increase their noise insulation performance (Huang, Qiu, and Kang, 2011).

For easy installation in practice, ANC systems can be applied directly on the entire openings. Ise (2005) installed 16 single-channel ANC systems on an open window using the boundary surface control principle, and achieved more than 10 dB attenuation at error locations from 200 Hz to 700 Hz. Seven years later, Murao and Nishimura (2012) proposed an active acoustic shielding (AAS) system, where four individual single-channel feedforward ANC units were installed in an open window, with the result that the measured noise reduction performance could be up to 2000 Hz in some local areas. The effects of multiple noise sources and moving noise sources were also investigated, and the AAS window was found to be able to have a large quiet area in a wide frequency range.

The physical limits of ANC open windows have been explored with finite element modeling in two dimensions, which revealed that the optimized number of secondary sources could be decided based on the required level of noise reduction (Lam et al., 2018). Elliott et al. (2018) investigated the relationship between the performance of the ANC system on openings and the configurations of secondary sources based on the wavenumber approach and the finite-element method. It is found that when the size of the window is compatible with the acoustic wavelength, only a few sources are needed for good control, but more secondary sources are needed for larger windows.

Section 2.5 discusses the theory for calculating the sound transmission via an opening into an enclosure, compares the performance of the surface and boundary control systems, and illustrates the upper-limit frequency of effective control of the systems.

### 2.5.1 Sound Transmission via an Opening into an Enclosure

Figure 2.18 shows the schematic diagram of a boundary control system where the secondary sources are installed evenly on the edge of the opening of a cavity with dimensions of $l_x \times l_y \times l_z$. All walls of the cavity are rigid except the opening, so the sound inside only comes from the opening. Points A to E in the figure are the locations of the primary source, and point P is at the center of the opening (Wang, Tao, and Qiu, 2019).

The sound pressure at location $\mathbf{r} = (x, y, z)$ inside the cavity from a primary point monopole source located at $\mathbf{r}_0 = (x_0, y_0, z_0)$ outside the cavity can be calculated similarly with the modal expansion method described in Section 2.4.2. When all walls inside the cavity are rigid, the sound pressure inside can be written as

$$p_{in}(x, y, z) = \sum_n \sum_m \left[ p_{n,m}^+ e^{-jk_{y,n,m}y} + p_{n,m}^- e^{jk_{y,n,m}y} \right] \cdot \phi_{n,m}(x, z) \qquad (2.98)$$

where $p_{n,m}^+$ and $p_{n,m}^-$ correspond to the $(n, m)$th mode in the positive and negative $y$ directions, $k_{y,n,m} = \sqrt{k^2 - (n\pi/l_x)^2 - (m\pi/l_z)^2}$ is the wavenumber in the $y$ direction, and $\phi_{n,m}(x, z) = \cos(n\pi x/l_x)\cos(m\pi z/l_z)$ is the eigenfunction of the $(n, m)$th mode.

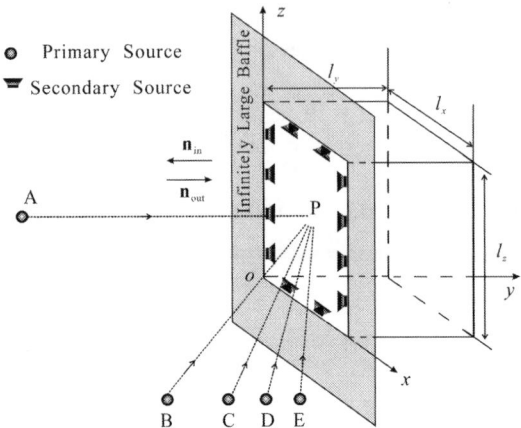

**FIGURE 2.18**
A virtual sound barrier system with the secondary loudspeakers located at the boundary of the opening for sound transmission control.

The boundary conditions at $y=0$ and $y=l_y$ are

$$\left.\frac{\partial p_{in}(x,y,z)}{\partial \mathbf{n}_{in}}\right|_{y=l_y} = 0 \tag{2.99}$$

$$p_{in}(x,y,z)\big|_{y=0} = p_{out}(x,y,z)\big|_{y=0} \tag{2.100}$$

$$\left.\frac{\partial p_{in}(x,y,z)}{\partial \mathbf{n}_{in}}\right|_{y=0} = \left.\frac{\partial p_{out}(x,y,z)}{\partial \mathbf{n}_{out}}\right|_{y=0} \tag{2.101}$$

where $\mathbf{n}_{in}$ and $\mathbf{n}_{out}$ are the direction vectors of the opening surface shown in Figure 2.18, and $p_{out}$ is the acoustic pressure outside the cavity, which can be obtained by summing the incident pressure $p_i$, the reflection pressure $p_r$, and the scattering pressure $p_{sca}$ as

$$p_{out}(x,y,z) = p_i(x,y,z) + p_r(x,y,z) + p_{sca}(x,y,z) \tag{2.102}$$

When the opening is on an infinitely large rigid baffle, the summation of the incident wave $p_i$ and the reflection pressure $p_r$ on the opening from the primary point monopole source can be obtained with the image source method (Kinsler et al., 2000)

$$p_i(x,0,z) + p_r(x,0,z) = \frac{2A}{r_0}e^{-jkr_0} \tag{2.103}$$

where $A$ in kg/s$^2$ is the amplitude of incident wave, $k$ is the wavenumber, and $r_0 = \sqrt{(x-x_0)^2 + y_0^2 + (z-z_0)^2}$. The scattering wave $p_{sca}$ can be calculated with the Rayleigh Integral by

$$p_{\text{sca}}(x,y,z) = \frac{1}{2\pi} \iint_S \frac{\partial p_{\text{out}}}{\partial n_{\text{out}}}\bigg|_{y=0} \left(\frac{e^{-jkr_{\text{sca}}}}{r_{\text{sca}}}\right) dS \tag{2.104}$$

where $S = l_x l_z$ stands for the opening surface, $r_{\text{sca}} = \sqrt{(x-x_s)^2 + y^2 + (z-z_s)^2}$.
Combining Equations (2.98) and (2.99) leads to

$$-p_{n,m}^+ e^{-jk_{y,n,m}l_y} + p_{n,m}^- e^{jk_{y,n,m}l_y} = 0 \tag{2.105}$$

The mode-matching equations are obtained according to the continuity of the sound pressure and particle velocity at the opening

$$F_{n,m} + \sum_p \sum_q k_{y,p,q}\left(-p_{n,m}^+ + p_{n,m}^-\right) \cdot \frac{S}{\rho_0\omega} Z_{n,m}^{p,q} = \left(p_{n,m}^+ + p_{n,m}^-\right)\varepsilon_n\varepsilon_m S \tag{2.106}$$

where

$$F_{n,m} = \int_0^{l_x}\int_0^{l_z} \frac{2A}{\sqrt{(x-x_0)^2 + y_0^2 + (z-z_0)^2}}$$
$$e^{-jk\sqrt{(x-x_0)^2 + y_0^2 + (z-z_0)^2}} \phi_{n,m}(x,z)\,dxdz. \tag{2.107}$$

$$Z_{n,m}^{p,q} = \frac{j\rho_0\omega}{2\pi S}\int_0^{l_x}\int_0^{l_z}\int_0^{l_x}\int_0^{l_z} \phi_{n,m}(x,z)\frac{e^{-jk\sqrt{(x-x_s)^2+(z-z_s)^2}}}{\sqrt{(x-x_s)^2+(z-z_s)^2}}$$
$$\phi_{p,q}(x_s,z_s)\,dxdzdx_sdz_s \tag{2.108}$$

with $\varepsilon_0 = 2$, and $\varepsilon_i = 1$ for $i \neq 0$, $\rho_0$ is the density of the air. $p_{n,m}^+$ and $p_{n,m}^-$ can be obtained by combining Equations (2.105) and (2.106) in matrix form as

$$\begin{pmatrix} \mathbf{P}^+ \\ \mathbf{P}^- \end{pmatrix} = \begin{pmatrix} \boldsymbol{\varphi}_1 & \boldsymbol{\varphi}_2 \\ \boldsymbol{\varphi}_3 & \boldsymbol{\varphi}_4 \end{pmatrix}^{-1} \begin{pmatrix} \mathbf{F} \\ \mathbf{0} \end{pmatrix} \tag{2.109}$$

with

$$\varphi_{1,n,m}^{p,q} = \varepsilon_n\varepsilon_m S\delta_{n,m}^{p,q} + \frac{S}{\rho_0\omega}k_{y,p,q}Z_{n,m}^{p,q}$$

$$\varphi_{2,n,m}^{p,q} = \varepsilon_n\varepsilon_m S\delta_{n,m}^{p,q} - \frac{S}{\rho_0\omega}k_{y,p,q}Z_{n,m}^{p,q} \tag{2.110}$$

$$\varphi_{3,n,m}^{p,q} = -e^{-jk_{y,p,q}l_y}\delta_{n,m}^{p,q}$$

$$\varphi_{4,n,m}^{p,q} = e^{jk_{y,p,q}l_y}\delta_{n,m}^{p,q}$$

where $\mathbf{P^+}$, $\mathbf{P^-}$ and $\mathbf{F}$ are $n \times m$ matrices, and the $(n, m)$th element are $p_{n,m}^+$, $p_{n,m}^-$, and $F_{n,m}$ respectively. The sound pressure at any location inside the cavity can then be calculated with Equation (2.98).

### 2.5.2 Control with Planar Virtual Sound Barriers

For the secondary sources installed on the boundary, the sound field can be obtained in a way similar to that described in Section 2.4.2,

$$
p_s(x,y,z) = \sum_n \sum_m \left[ p_{s,n,m}^+ e^{-jk_{y,n,m}y} + p_{s,n,m}^- e^{jk_{y,n,m}y} \right] \cdot \phi_{n,m}(x,z)
$$
$$
+ \iiint_V j\rho_0\omega q(\mathbf{r}_s) G_A dV \tag{2.111}
$$

where $p_s(x, y, z)$ is the sound pressure at point $(x, y, z)$ in the cavity, $p_{s,n,m}^+$ and $p_{s,n,m}^-$ represent the parameters of the $(n, m)$th mode for the positive and negative directions respectively, and $\mathbf{r}_s$ is the location of the secondary source. $p_{s,n,m}^+$ and $p_{s,n,m}^-$ can be obtained in a way similar to that for the primary source by using the boundary conditions, and the Green's function, $G_A$, is given in Equation (2.73).

The cost function for optimizing the secondary source strength is selected as

$$
J = \frac{1}{2\rho_0 c_0^2} |p_t|^2 + \beta \mathbf{q}_s^H \mathbf{q}_s \tag{2.112}
$$

where $p_t$ is the total sound pressure at evaluation points with control, $\beta$ is a positive real number for constraining the control effort, and $\mathbf{q}_s$ in m³/s are the strengths of the secondary sources. The optimized strengths of the secondary sources are

$$
\mathbf{q}_s = -\frac{1}{2\rho_0 c_0^2} \left( \frac{1}{2\rho_0 c_0^2} \mathbf{Z}_{se}^H \mathbf{Z}_{se} + \beta \mathbf{I} \right)^{-1} \mathbf{Z}_{se}^H \mathbf{Z}_{pe} q_p \tag{2.113}
$$

where $\mathbf{Z}_{se}$ is the transfer function from the $M$ secondary sources to the $L$ error points, $\mathbf{Z}_{pe}$ is the transfer function from the primary source to the $L$ error points, and $q_p$ is the source strength of the primary source. $\mathbf{Z}_{pe}$ and $\mathbf{Z}_{se}$ in Pa·s/m³ can be obtained by the theoretical calculation, simulations, and experimental measurements.

The performance of the boundary control system is evaluated with the ratio of the summation of the squared sound pressure in the target area without and with control,

$$\mathrm{NR} = 10\log_{10}\dfrac{\displaystyle\sum_{i=1}^{N_\mathrm{v}}|p_\mathrm{p}(\mathbf{r}_{\mathrm{v},i})|^2}{\displaystyle\sum_{i=1}^{N_\mathrm{v}}|p_\mathrm{t,o}(\mathbf{r}_{\mathrm{v},i})|^2} \qquad (2.114)$$

where $\mathbf{r}_{\mathrm{v},i}$, $i=1$, 2, ..., $N_\mathrm{v}$, are the locations of the evaluation points, and $N_\mathrm{v}$ is the number of evaluation points, which is chosen to ensure at least 6 evaluation points per wavelength.

The performances of the boundary control system for an open cavity (0.432 m×0.598 m×0.67 m) shown in Figure 2.18 are compared with the surface control one, which distributes 12 secondary sources evenly on the opening. All secondary sources are on the plane of $y=0.1$ m, and the distributions of secondary sources are shown in Figure 2.19, where (a), (b), and (c) are for the boundary control distribution of 4, 8, and 12 secondary sources, respectively, and (d) is for the surface control. There are 22×22×22 error points uniformly distributed in the cavities for calculating the cost function, while another 24×24×24 evaluation points uniformly distributed in the cavities (at least 6 evaluation points in one wavelength at 2000 Hz) are used for evaluating the system performance. And there are 100 modes used in the calculation in order to ensure the accuracy for frequencies below 2000 Hz.

The normal incidence case is approximated by locating the primary source at (0.216, −2.0, 0.335) m (point A in the Figure 2.18). The averaged sound pressure level without and with control and the noise reduction are shown in Figure 2.20, where the noise reduction is high at the cavity resonance frequencies but low at the other frequencies when the primary sound pressure level is low. The performance of the surface control system is a little better than that of the boundary control system but at the price of having secondary sources in the sound propagation path. The upper-limit frequencies (the lowest frequency when NR is less than 10 dB) are about 500 Hz for the three configurations of the boundary control systems, which is lower than the frequency (787 Hz) corresponding to $kl_x=2\pi$ ($l_x=0.432$ m). The $NR$ curves indicate that adding more secondary sources (from 4 to 12) on the boundary does not give a significant improvement. The upper-limit frequency of the control for this cavity is lower than that with the same size opening (0.432 m×0.67 m) in free space.

### 2.5.3 The Upper-Limit Frequency

The noise reduction performances with different cavity depths of 0.299 m, 0.598 m, 1.196 m, 2.392 m, and 4.784 m are calculated for the boundary control system. The averaged sound pressure levels without and with control and the noise reductions are shown in Figure 2.21, where the noise reduction is higher when the depth is larger. The $kl_x$ corresponding to the upper-limit frequency for a depth from 0.299 m to 4.784 m are 0.9π, 1.3π, 1.7π, 1,9π, and 1.9π, respectively. It seems that the upper-limit frequency can be roughly estimated by

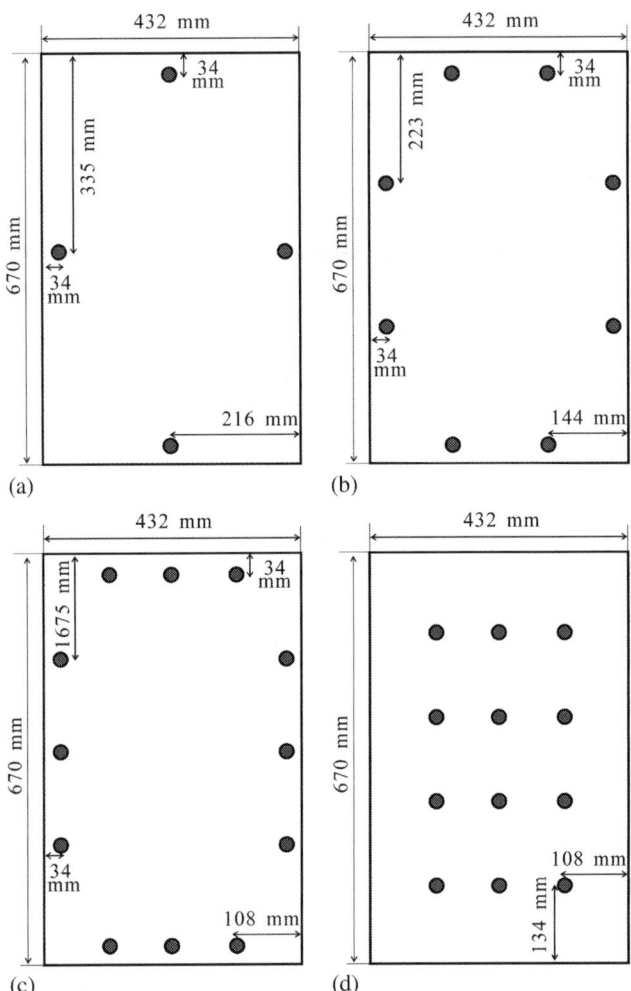

**FIGURE 2.19**
Distributions of the secondary sources on the plane of $y=0.1$ m: (a) 4 secondary sources on the boundary, (b) 8 secondary sources on the boundary, (c) 12 secondary sources on the boundary, and (d) 12 secondary sources distributed evenly on the opening surface.

$f_{max}=c_0/w$, i.e., $kw=2\pi$, where $w$ is the width of the opening for the cavities with sufficiently large depth. This is consistent with the results in free space.

For a primary point monopole source located at $(-0.5, -0.5, -0.5)$ m corresponding to the point C in Figure 2.18 (the incident angle to the opening surface is 51°), the performances are shown in Figure 2.22. The upper-limit frequencies are 453 Hz, 688 Hz, and 961 Hz for 4, 8, and 12 secondary sources, which are not limited by $kl_x=2\pi$ (787 Hz). This is consistent with the result in free space, but the estimation for the upper-limit frequency for this case becomes more complicated because the depth of the cavity is also a factor.

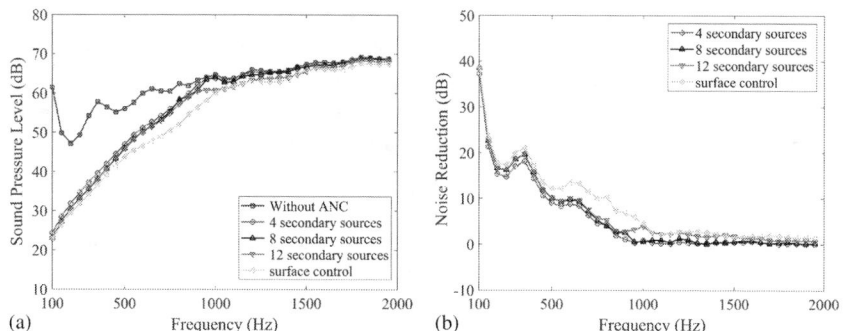

**FIGURE 2.20**
The noise reduction performance under normal incidence of the virtual sound barrier systems: (a) the averaged sound pressure levels without and with control, and (b) the noise reductions for these 4 configurations.

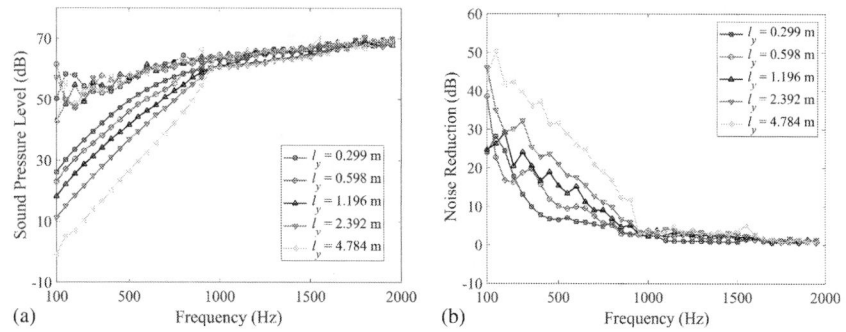

**FIGURE 2.21**
The noise reduction performance of the boundary control system with 12 secondary sources for different depths under normal incidence: (a) the averaged sound pressure levels without (dashed lines) and with (solid lines) control, and (b) the noise reductions for these 5 configurations.

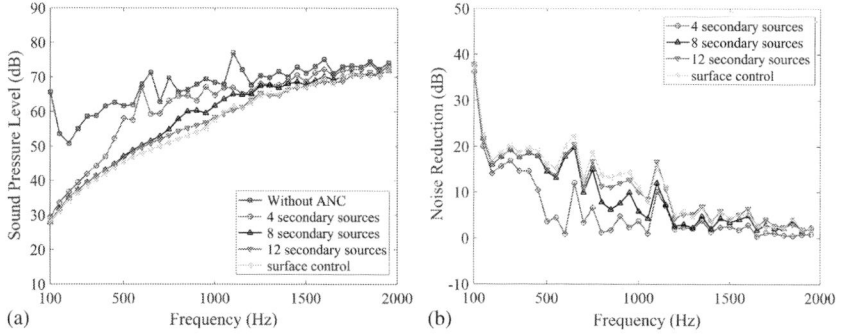

**FIGURE 2.22**
The noise reduction performance of the virtual sound systems with 4 different secondary source distributions: (a) the averaged sound pressure levels without and with control, and (b) the noise reductions for a cavity with depth of 0.598 m.

In Figure 2.18, positions B, C, D, and E are at (−0.5, −1.0, −0.5) m, (−0.5, −0.5, −0.5) m, (−0.5, −0.25, −0.5) m, and (−0.5, −0.03, −0.5) m, respectively, which correspond to the incident angles to the opening surface of 51°, 68°, 78°, and 89°. The NR of the boundary control system with 12 secondary sources for the cavity with a depth of 0.598 m in these 4 situations and the normal incidence case discussed above are shown in Figure 2.23. The NR of normal incidence is lower than that of the other 4 incident angles, and the NR does not vary significantly with large incident angles (from 51° in the simulations).

In summary, the surface control virtual sound barrier systems can have broad band control up to a high frequency as long as the space between the secondary sources is sufficiently small. It is also feasible to apply the boundary control system for sound transmission control through an opening into a cavity, which only installs the secondary sources on the frame of the opening and thus has the advantages of better ventilation and easier access. The boundary control system can effectively control sound transmission through the opening up to a certain frequency. This effective upper-limit frequency is decided by the length of the short sides of the opening, which is equal to the wavelength of this frequency. The performance of the boundary control system can be greater than that of the surface control system when the number of secondary sources is limited or when the opening is narrow. The performance of the boundary control system does not vary significantly for oblique incidence with large incident angles.

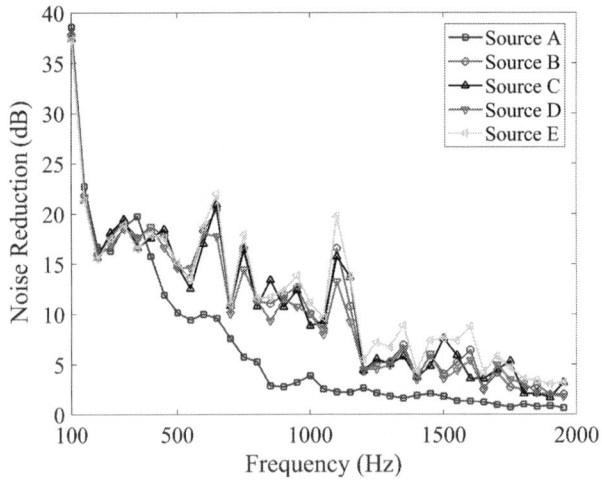

**FIGURE 2.23**
The noise reductions of the boundary control system with 12 secondary sources for 5 primary source locations.

# 3

## Three-Dimensional Virtual Sound Barriers

### 3.1 Problem Description

In enclosed spaces, or sometimes in open spaces, noise comes from many different directions due to multiple reflections from the boundaries of the enclosures or from noise sources at different locations. In these situations, three-dimensional virtual sound barriers are desired because planar virtual sound barriers can only block sound propagation from a single direction, or only a few directions. Figure 3.1 illustrates the problem to be studied, in which an array of loudspeakers are located in a three-dimensional closed form to create a quiet zone within the space surrounded by the error sensors in a noisy environment.

The monitoring sensors can be omitted for a steady primary noise field, where the radiation of the loudspeakers can be predesigned in advance according to the characteristics of the primary noise field, the location of

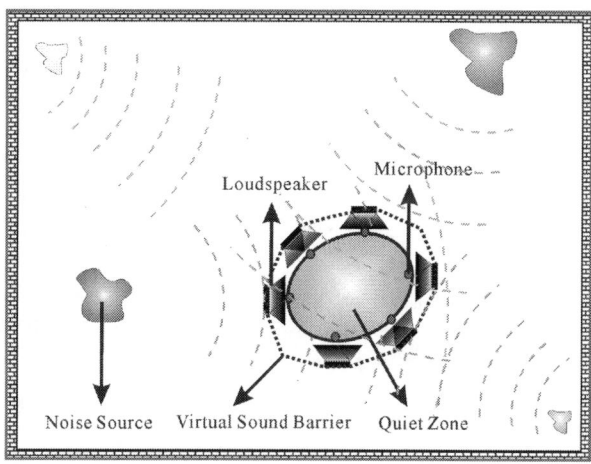

**FIGURE 3.1**
Schematic drawing of a three-dimensional virtual sound barrier.

the quiet zone, and the arrangement of the loudspeakers. For a general primary noise field where the position, the amplitude, and the frequency of the noise sources are time varying, good noise reduction performance of the systems can be achieved with an adaptive controller that adjusts the output signals to the loudspeakers by using the input signals of the monitoring sensors.

The control mechanisms of the virtual sound barrier systems might differ, according to the type of the secondary sources and the control strategies. Since a virtual sound barrier system is usually far away from the primary source, the mechanism is generally an absorption or reflection of the sound energy of the primary source, rather than reducing the impedance seen by the primary source; however, for the virtual sound barrier systems inside an enclosure with strong resonance at the dominant modes, the mechanism can be an unloading of the impedance seen by the primary source via modal coupling.

## 3.2 Creation of a Quiet Zone in a Noisy Environment

This section introduces a model for describing a noise field that has sound from many directions, and then provides formulae to carry out simulations and designs for three-dimensional virtual sound barrier systems. Finally, simulation and experiment results are presented to illustrate their basic properties.

### 3.2.1 Formulation

The primary sound field at point $\mathbf{r}$ is assumed to consist of a number of plane waves with random phases from many different directions as

$$p_p(\mathbf{r}) = \frac{1}{\sqrt{N_p}} \sum_{i=1}^{N_p} A_{p,i} e^{-j(k\mathbf{n}_i \cdot \mathbf{r} + \varphi_{0,i})} \tag{3.1}$$

where $N_p$ is the number of plane waves, and the $i$th plane wave with magnitude of $A_{p,i}$ comes from a random direction $\mathbf{n}_i$ with a random phase $\varphi_{0,i}$. $\mathbf{n}_i = (n_x, n_y, n_z)$ is a direction vector with each of its components being uniformly distributed in $(-1, 1)$. The magnitude $A_{p,i}$ and the phase $\varphi_{0,i}$ are taken from uniform distributions of $(0, A_0)$ and $(-\pi, \pi)$, respectively. $k = \omega/c_0$ is the wave number, $c_0$ is the speed of sound in the air, $\omega = 2\pi f$ is the angular frequency, and $f$ is the frequency. When $N_p$ is sufficiently large, the primary sound field given above can be used to approximate a diffuse sound field (Nelson and Elliott, 1992).

Assume the secondary sound sources are point monopoles and only the direct sound is taken into account in the simulations for simplicity. The secondary sound field generated by the virtual sound barrier is the superposition of $N_s$ secondary monopole sources, which can be expressed by

$$p_s(\mathbf{r}) = \sum_{i=1}^{N_s} \frac{j\omega\rho_0 q_{s,i}}{4\pi|\mathbf{r} - \mathbf{r}_{s,i}|} e^{-jk|\mathbf{r}-\mathbf{r}_{s,i}|} \tag{3.2}$$

where $q_{s,i}$ is the complex source strength of the $i$th secondary source located at $\mathbf{r}_{s,i}$. The total sound pressure at location $\mathbf{r}$ can be calculated by

$$p_t(\mathbf{r}) = \frac{1}{\sqrt{N_p}} \sum_{i=1}^{N_p} A_{p,i} e^{-j(k\mathbf{n}_i \cdot \mathbf{r}+\varphi_{0,i})} + \sum_{i=1}^{N_s} \frac{j\omega\rho_0 q_{s,i}}{4\pi|\mathbf{r} - \mathbf{r}_{s,i}|} e^{-jk|\mathbf{r}-\mathbf{r}_{s,i}|} \tag{3.3}$$

In matrix form, this can be expressed as

$$p_t(\mathbf{r}) = p_p(\mathbf{r}) + \mathbf{Z}_s(\mathbf{r})\mathbf{q}_s \tag{3.4}$$

with $\mathbf{q}_s = [q_{s,1}, q_{s,2}, \ldots, q_{s,Ns}]^T$, $\mathbf{Z}_s(\mathbf{r}) = [Z_{s,1}(\mathbf{r}), Z_{s,2}(\mathbf{r}), \ldots, Z_{s,Ns}(\mathbf{r})]$, and the $i$th element of $\mathbf{Z}_s(\mathbf{r})$ is,

$$Z_{s,i}(\mathbf{r}) = \frac{j\omega\rho_0}{4\pi|\mathbf{r} - \mathbf{r}_{s,i}|} e^{-jk|\mathbf{r}-\mathbf{r}_{s,i}|} \tag{3.5}$$

The cost function to be minimized is defined as

$$J = \mathbf{p}_t^H \mathbf{p}_t + \beta \mathbf{q}_s^H \mathbf{q}_s \tag{3.6}$$

where $\mathbf{p}_t = [p_t(\mathbf{r}_{e,1}), p_t(\mathbf{r}_{e,2}), \ldots, p_t(\mathbf{r}_{e,Ne})]^T$, $N_e$ is the number of error sensors located at $\mathbf{r}_{e,i}$ and $\beta$ is a positive real number being used to determine the weighting for the control effort term. Substituting Equation (3.4) into Equation (3.6), gives

$$J = \mathbf{q}_s^H (\mathbf{A} + \beta\mathbf{I})\mathbf{q}_s + \mathbf{q}_s^H \mathbf{b} + \mathbf{b}^H \mathbf{q}_s + c \tag{3.7}$$

where

$$\mathbf{A} = \mathbf{Z}_{se}^H \mathbf{Z}_{se}, \quad \mathbf{b} = \mathbf{Z}_{se}^H \mathbf{p}_p, \quad c = \mathbf{p}_p^H \mathbf{p}_p \tag{3.8}$$

with $\mathbf{Z}_{se} = [\mathbf{Z}_s(\mathbf{r}_{e,1})^T, \mathbf{Z}_s(\mathbf{r}_{e,2})^T, \ldots, \mathbf{Z}_s(\mathbf{r}_{e,Ne})^T]^T$, and $\mathbf{p}_p = [p_p(\mathbf{r}_{e,1}), p_p(\mathbf{r}_{e,2}), \ldots, p_p(\mathbf{r}_{e,Ne})]^T$. The optimal vector of the secondary source strength is given by

$$\mathbf{q}_s = -(\mathbf{A} + \beta\mathbf{I})^{-1}\mathbf{b} \tag{3.9}$$

Once the optimal secondary source strength has been obtained, it can be substituted back into Equation (3.3) to calculate the total sound pressure amplitude with control, which is denoted as $p_{t,o}$. The performance of the virtual sound barrier system considered in this section is defined as the ratio of the sum of the squared sound pressure inside a defined volume without and with control, i.e., the noise reduction (NR), as

$$\text{NR} = 10\log_{10} \frac{\sum_{i=1}^{N_v} |p_\text{p}(\mathbf{r}_{v,i})|^2}{\sum_{i=1}^{N_v} |p_{t,o}(\mathbf{r}_{v,i})|^2} \tag{3.10}$$

where $\mathbf{r}_{v,i}$, $i=1, 2, ..., N_v$, are the locations of the evaluation points, and $N_v$ is the number of evaluation points, which is chosen to ensure at least 6 evaluation points per wavelength.

### 3.2.2 Two-Dimensional Simulations

The two-dimensional sound field considered first is where the primary sound field consists of the plane waves travelling on the $x$–$y$ plane only. It can be calculated by forcing $n_z=0$ in Equation (3.1). With this assumption, the secondary source array and the error sensor array can be in the shape of a ring on the $x$–$y$ plane, centered at the origin of the coordinate systems shown in Figure 3.2.

The amplitude and phase distribution of the primary sound field at 1500 Hz are calculated with $A_0=1.0$ and $N_p=100$, and the results of one trial are shown in Figure 3.3 for the sound pressure distribution on the $x$–$y$ plane in a square of 2.0 m$\times$2.0 m, centered at the origin of the coordinate. The darkest color for the amplitude is approximately 1.5 and the lightest color is for 0.0. The phase is from 180° to –180°. The sound field appears quite complicated with no obvious pattern.

In the simulations, the control force weighting parameter $\beta$ in the cost function of Equation (3.6) is set at 0.01 of the largest eigenvalue of the corresponding matrix **A**. Although this value of $\beta$ may not obtain the maximum noise reduction in some cases, the simulations show that stable and consistent performance can be maintained under various conditions for such a value of $\beta$.

The system performance considered in this section is defined as the ratio of the sum of the squared sound pressure inside a 0.15 m high cylinder with a diameter of 0.3 m without and with control. The cylinder is centered at the origin of the coordinates and the height is along the z-axis. To show the influence of the virtual sound barrier system on the other areas of the room, another index called TR (total reduction) is also calculated in a similar way within a cube with dimensions of 2 m × 2 m × 2 m centered at the origin. The space between two evaluation points is 0.05 m, which

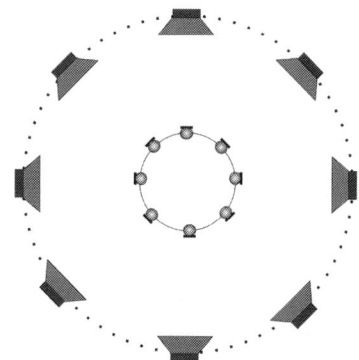

**FIGURE 3.2**
Circular secondary source array with a diameter of 1.0 m and a circular error sensor array with a diameter of 0.3 m on the *x–y* plane.

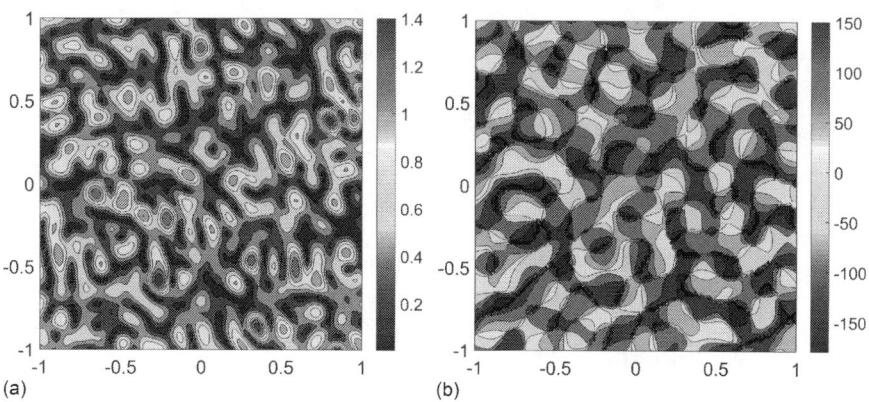

(a)                                                                                    (b)

**FIGURE 3.3**
An example of the distribution of the primary sound field on the *x–y* plane at 1500 Hz: (a) amplitude, (b) phase.

provides sufficient precision for sound up to 1500 Hz. As point monopoles are used as the secondary sources and the sound pressure near the secondary sources tends to be very large, the sound pressure at the points within a distance of 0.05 m from the secondary sources are removed from the calculation of NR and TR.

The perimeter of the evaluation cylinder on the *x–y* plane is about 0.94 m, and half of the corresponding acoustical wavelength at 1500 Hz is approximately 0.12 m, so 8 error sensors and secondary sources are used in the simulations, as shown in Figure 3.2. Figure 3.4 shows the sound pressure distribution with control in three planes of the evaluation cube when the direction of the primary plane wave is $\mathbf{n}_i = (1, 0, 0)$. In the figure, the darkest color corresponds to the sound pressure amplitude of 2.0 (for sound pressure greater than 2.0, it is clipped at 2.0 in the drawing) while the lightest color corresponds to zero amplitude. The primary sound pressure amplitude is 1.0, and the diameters of the circular secondary source array and the error sensor array rings are 1.0 m and 0.3 m, respectively. The corresponding NR is 9.1 dB, and TR is −1.2 dB, which means a small increase of the sound energy outside the evaluation cylinder.

Figure 3.5 shows NR as a function of the propagation direction of the primary noise from 0° to 90° at 1500 Hz. The NR is dependent on the propagation direction of the primary noise due to the locations of the secondary sources. The curves in the figure are for the systems with 4, 8, and 16 channels, respectively (from low noise reduction to high). For example, in an 8-channel system, 8 secondary sources and 8 error sensors are uniformly distributed on the perimeters of 2 rings with diameters of 1.0 m and 0.3 m, respectively. For the 8-channel system, NR achieves the maximum at the

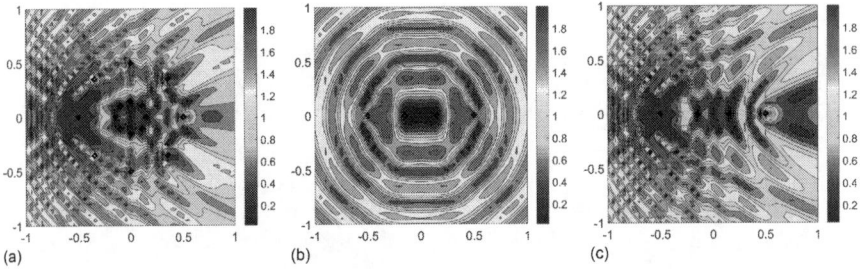

**FIGURE 3.4**
Sound pressure amplitude distribution on different planes of the virtual sound barrier system, with the diameters of the error sensor ring and secondary source rings being 0.3 m and 1.0 m, respectively: (a) $x$–$y$ plane, (b) $y$–$z$ plane, (c) $x$–$z$ plane.

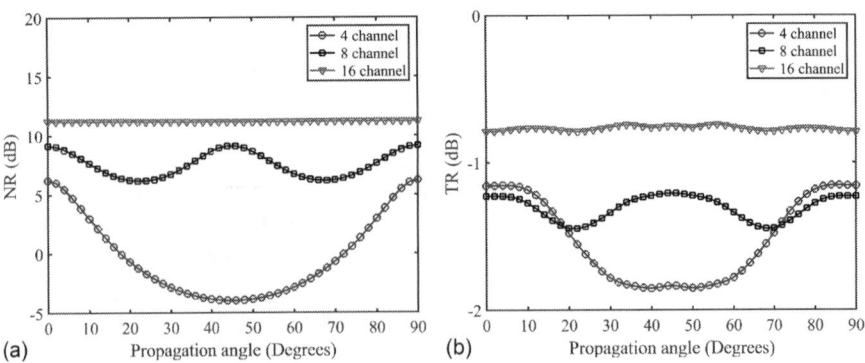

**FIGURE 3.5**
(a) NR and (b) TR as a function of the propagation direction of the primary noise for the system, with the diameters of the error sensor ring and secondary source rings being 0.3 m and 1.0 m, respectively; the curves (from low noise reduction to high) in the figure are for systems with 4-, 8- and 16-channels, respectively.

angles of 0°, 45°, and 90° and the minimum at 22.5° and 67.5°. This is not surprising if one refers to Figure 3.2. When there is a secondary source in the direction of primary sound propagation, the control sound field is easy to match the primary sound field and thus achieves high attenuation. At the angles of 22.5° and 67.5°, there is no secondary source located in this direction, thus less noise reduction is obtained. For the primary sound field consisting of a number of plane waves from different directions with random phase, the NR for the above secondary source and error sensor configuration is a value between the maximum and the minimum values in Figure 3.5.

The diameter of the secondary source ring affects the performance of the system. For an 8-channel system with the primary sound field consisting of just one plane wave propagating at 22.5 degrees, Figure 3.6 shows the NR and TR with different diameters of the secondary source ring at 250 Hz, 500 Hz, 1000 Hz, and 1500 Hz, respectively. In the simulations, the diameter of the error sensor ring is fixed at 0.3 m. It can be observed that the noise reduction of the system is higher at a lower frequency, and the noise reduction of the system increases with the diameter of the secondary source ring. This is because the control sound field is more like a plane wave at a larger distance from the error sensors. However, its side effect is that there is usually a little sound energy increase outside the evaluation cylinder, as illustrated in Figure 3.6(b).

The diameter of the error sensor ring also affects the performance of the system. For the 8-channel system with the primary sound field consisting of just one plane wave propagating at 22.5 degrees, Figure 3.7 shows the NR and TR with different diameters of the error sensor ring at 250 Hz, 500 Hz, 1000 Hz, and 1500 Hz, respectively. In the simulations, the diameter of the secondary source ring is fixed at 2.0 m. It can be seen that there is an optimum diameter for the error sensor ring, which is not larger than the wavelength under consideration. For very small diameter of the error sensor ring,

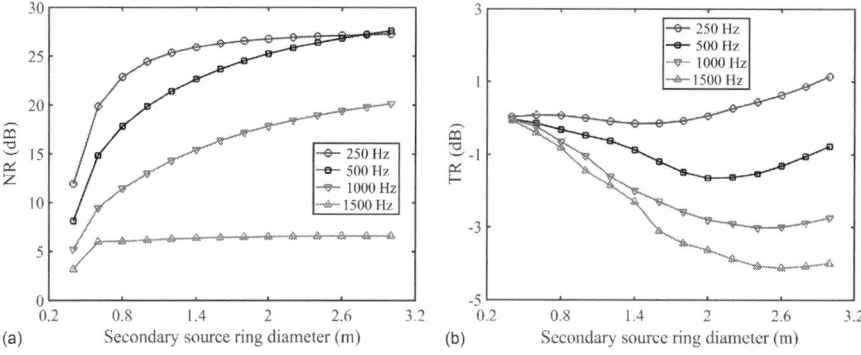

**FIGURE 3.6**

(a) NR and (b) TR as a function of the diameter of the secondary source ring for the system, with the diameters of the error sensor ring being 0.3 m; the curves (from high noise reduction to low) correspond to those at 250 Hz, 500 Hz, 1000 Hz, and 1500 Hz, respectively.

the sound field inside the ring does reduce; however, because the evaluation cylinder is much larger than the error sensor ring, the total noise reduction in the evaluation cylinder decreases. When the diameter of the error sensor ring is larger than half of the sound wavelength or the diameter of the evaluation zone, the noise reduction inside the evaluation cylinder decreases.

Figure 3.8 shows the NR and TR as a function of frequency from 100 Hz to 5000 Hz for a primary plane wave propagating at 22.5 degrees. The curves in the figure are for the systems with 4, 8, and 16 channels respectively (from low noise reduction to a high one). It can be observed that the effective frequencies with positive noise reductions for the systems with 4, 8, and 16 channels are approximately 900 Hz, 1700 Hz, and 5100 Hz. Therefore, more channels can provide larger noise control for a broader band.

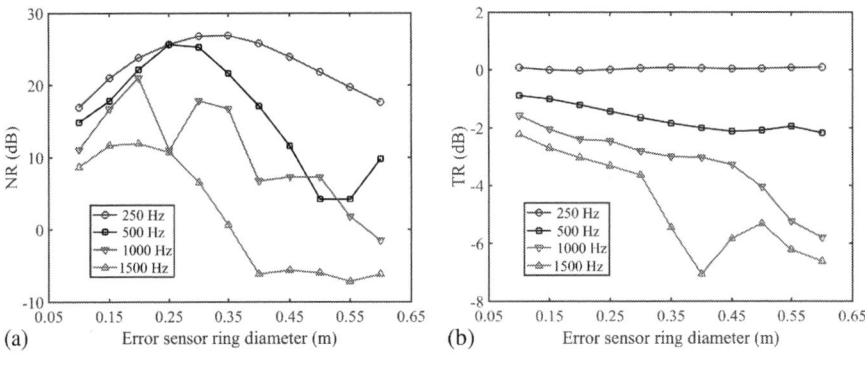

**FIGURE 3.7**
(a) NR and (b) TR as a function of the diameter of the error sensor ring for the system, with the diameter of the secondary source ring being 2.0 m; the curves (from high noise reduction to low) correspond to those at 250 Hz, 500 Hz, 1000 Hz, and 1500 Hz, respectively.

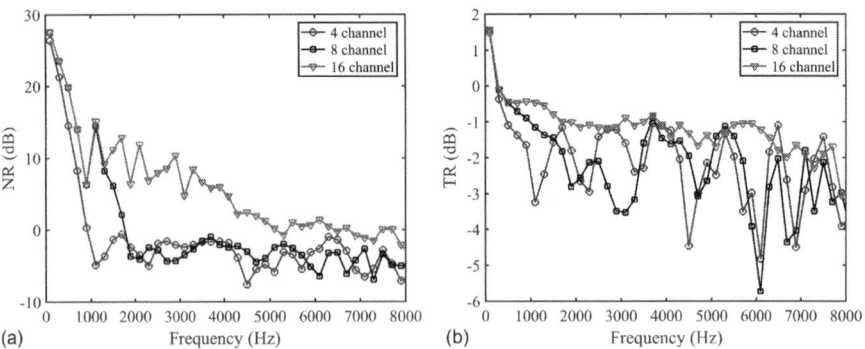

**FIGURE 3.8**
(a) NR and (b) TR as a function of the frequency of the system, with the diameters of the error sensor ring and secondary source rings being 0.3 m and 1.0 m, respectively, and with the primary sound field consisting of just one plane wave propagating at 22.5 degrees; the curves (from low noise reduction to high) in the figure are for the systems with 4-, 8- and 16-channels, respectively.

All these two-dimensional sound field simulations with a circular secondary source array and a circular error sensor array reveal some fundamentals for applying the virtual sound barrier system. First, it is possible to generate a reasonably sizes quiet zone (larger than a human head) up to the middle audio frequency (for example, 1500 Hz) with a reasonable number of secondary sources and error sensors (8 secondary sources and error sensors). Second, larger and broader band attenuation can be obtained with more channels. For the primary sound field with many plane waves, better performance in the evaluation cylinder can be obtained when the secondary sources are further away from the error sensors, but at the cost of introducing sound energy at the other locations. The locations of the error sensors are also important, and they should be inside the evaluation cylinder, but with a sufficiently large occupation area.

### 3.2.3 Three-Dimensional Simulations

It is shown in the previous section that it is possible to use eight secondary point monopole sources to generate a quiet zone of a useful size if the primary sound comes only from $x$ and $y$ directions. However, this kind of situation is rarely encountered in practice, because it requires not only that the ceiling and floor are completely absorptive but also that the primary sound sources are in the same horizon plane as the secondary sources. When there is a plane wave coming from the $z$ direction, the performance of the configuration mentioned in the previous section deteriorates dramatically. For example, if the primary sound field consists of just one plane wave propagating at direction $n_i = (0, 0, 1)$, the NR at 1500 Hz becomes $-6.9$ dB for the system, with the diameters of the secondary source ring and the error sensor ring on the horizon plane being 1.0 m and 0.15 m, respectively. There is an increase of the sound pressure level inside the evaluation cylinder although the sum of the squared sound pressure at the error sensors is still minimized.

To cope with the primary plane wave coming from the $z$ direction, the secondary sources need to be located in the $z$ direction as well. Figure 3.9 shows such a configuration, where the secondary point monopole sources are distributed nearly evenly on the sphere with a diameter of 1.0 m. They are spaced with an elevation of every 45 degrees and the number of the secondary sources on a horizontal ring is 8, also spaced evenly at an angle of 45 degrees, so the total number of the secondary sources is 26. The error sensors are located similarly inside on a sphere with a diameter of 0.15 m. The evaluation cylinder is still the same as that in Section 3.2.2, but the control force weighting parameter $\beta$ in the cost function of Equation (3.6) is set at 0.0001 of the largest eigenvalue of the corresponding matrix $\mathbf{A}$ in the three-dimensional simulations.

Figure 3.10 shows the sound pressure distribution with control in three planes of the evaluation cube when the direction of a unit magnitude primary plane wave is $n_i = (0, 0, 1)$. It can be seen that the actual 10 dB quiet zone

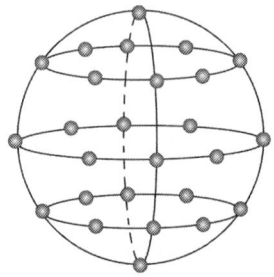

**FIGURE 3.9**
Configuration of a three-dimensional virtual sound barrier with a spherical array of the secondary sources as indicated by the small circles on the sphere surface.

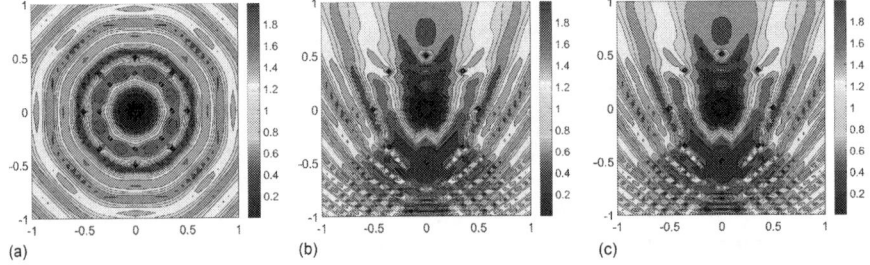

**FIGURE 3.10**
Sound pressure amplitude distribution on different planes of the system with the diameters of the error sensor sphere and the secondary source sphere being 0.15 m and 1.0 m, respectively, for a primary plane wave from $n_i = (0, 0, 1)$: (a) $x$–$y$ plane, (b) $y$–$z$ plane, (c) $x$–$z$ plane.

at 1500 Hz can be a sphere with a diameter of 0.42 m, larger than the evaluation cylinder. The NR is 20.0 dB and TR is –1.4 dB.

The NR for the spherical secondary source array is also dependent on the angle of incidence. The minimum NR is 10.6 dB when the direction of the primary plane wave is $n_i = (\cos\theta\cos\varphi, \cos\theta\sin\varphi, \sin\theta)$ with both $\theta$ and $\varphi$ being 22.5 degrees. For a primary sound field consisting of a number of plane waves from many different directions with random phase, NR for the above secondary source and error sensor configuration is a value between the maximum and the minimum values. Figure 3.11 shows the corresponding sound pressure distribution with control on three planes of the evaluation cube.

### 3.2.4 The 2.5-Dimensional Simulations

As sometimes it is not practical to place secondary sources at the locations corresponding to the directions of ceiling and ground, a cylindrical secondary source array can be used. Figure 3.12 shows a configuration where three

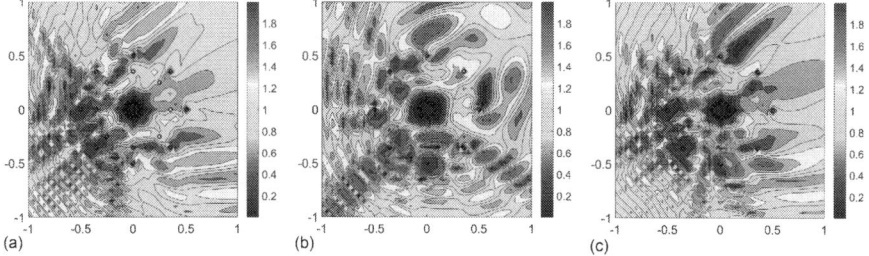

**FIGURE 3.11**
Sound pressure amplitude distribution on different planes of the system, with the diameters of the error sensor sphere and the secondary source sphere being 0.15 m and 1.0 m, respectively, for a primary plane wave from $n_i = (\cos45°\cos45°, \cos45°\sin45°, \sin45°)$: (a) $x$–$y$ plane; (b) $y$–$z$ plane; (c) $x$–$z$ plane.

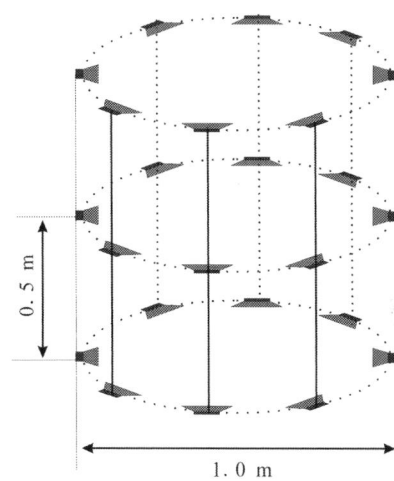

**FIGURE 3.12**
Configuration of a cylindrical virtual sound barrier with an array of secondary sources on a cylinder.

layers of secondary source rings are used. The diameter of the secondary source ring is 1.0 m and the space between the rings is 0.5 m, so the total height of the secondary source cylinder is 1.0 m. There are 8 point monopole sources distributed evenly on each ring, thus the total number of secondary sources is 24. The error sensor setup remains the same as that in Section 3.2.3, where they are on a sphere with a diameter of 0.15 m. The evaluation cylinder is also the same as that in Section 3.2.3, but the control force weighting parameter $\beta$ in the cost function of Equation (3.6) is set at 0.001 of the largest eigenvalue of the corresponding matrix $\mathbf{A}$ in the simulations.

The NR for the primary sound field consisting of just one plane wave from $n_i = (0, 0, 1)$ is only 2.9 dB, which is the worst case for this configuration, because there is no secondary source corresponding to the sound from the top direction. Fortunately, in a normal room, the primary sound consists of many plane waves from many directions, so the noise reduction performance depends on the proportion of the plane waves from different directions. For the primary sound field, consisting of 100 plane waves from different directions (random, uniformly distributed) with random phase, the NR is about 13.8 dB, while the maximum and minimum NRs are 22.1 dB and 2.9 dB, respectively.

Figure 3.13 shows a more practical setup with the error sensors. The setup of the secondary sources is the same as that in Figure 3.12, but the number of error sensors is reduced to 18, consisting of two disks with a diameter of 0.15 m, separated by 0.07 m. In each disk, there are 8 error sensors distributed evenly on the perimeter and 1 in the center.

The NRs for the noise from $(1, 0, 0)$ and $(0, 0, 1)$ directions are 14.9 dB and 2.7 dB respectively. The NR for a random incidence primary wave, consisting of 100 plane waves from different directions (random, uniformly distributed)

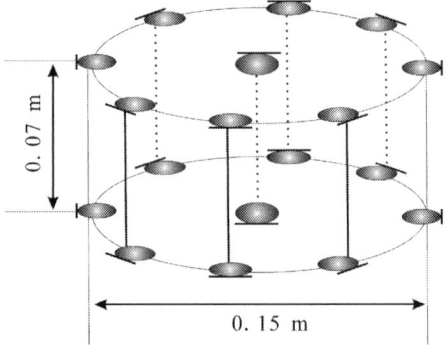

**FIGURE 3.13**
Configuration of a cylindrical error sensor array.

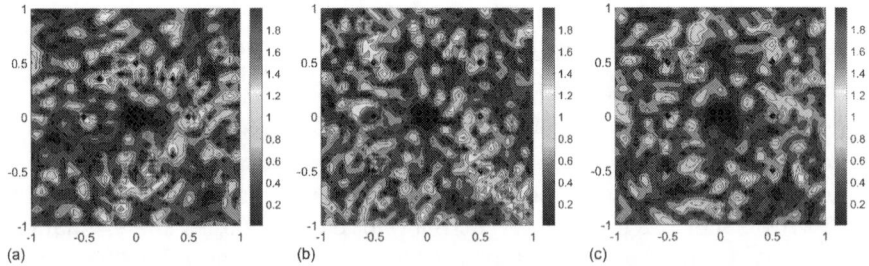

**FIGURE 3.14**
Sound pressure amplitude distribution on different planes of a system with the diameters of the error sensor cylinder and the secondary source cylinder being 0.15 m and 1.0 m, respectively, for a random incidence primary wave: (a) $x$–$y$ plane, (b) $y$–$z$ plane, (c) $x$–$z$ plane.

with random phase, is 11.0 dB. Figure 3.14 shows the sound field distributions with control, where the quiet area looks like a cylinder. All these simulation results demonstrate that it is feasible to generate a quiet zone of a useful size at a higher frequency with reasonable numbers of secondary sources and error sensors.

### 3.2.5 Experiments

The control performance of the 2.5-dimensional system discussed in Section 3.2.4 was experimentally investigated with a 16-channel cylindrical system (Zou et al., 2007). As shown in Figure 3.15, 16 error sensors were placed on two horizontal planes separated by $h_e$, and the 8 error sensors on each plane were evenly spaced in a circle with a radius of $r_e = h_e$; the secondary sources were located similarly surrounding the error sensors in two circles with a radius of $r_c = h_c$ on two horizontal planes separated by $h_c$. In the simulations for this configuration, the number of plane waves in Equation (3.1) is set to $N_p = 100$, and $\beta$ is set to 0.001 of the largest eigenvalue of the corresponding matrix so that the pressure attenuation observed at the error sensors is less than 40 dB in the simulations. The evaluation cylinder is the same as before (Zou et al., 2007).

The primary sound sources used in the experiments were three loud-speakers located in three different directions, at a height of 1.2 m, 3.0 m, and 0.5 m, respectively. Their inputs were from an amplified tonal signal. With the reflections of all the surfaces of the room, the primary sound field in the room was quite complicated. The virtual sound barrier system was placed in the center of the room with the central horizontal plane 0.8 m above the floor. A 16-channel active noise control (ANC) controller using the filtered reference LMS (Least Mean Squares) algorithm was applied. The tonal signal generated by a signal generator was also fed into the controller as the

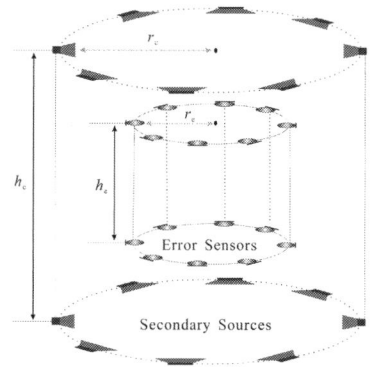

**FIGURE 3.15**
Configuration of a 16-channel cylindrical virtual sound barrier system.

reference signal. The distance between each primary source and the center of the system is about 4 m. Additional monitoring microphones were used to measure the sound pressure in the target region, with the intervals of the monitoring microphones less than 1/6 wavelength of the sound under investigation. Figure 3.16 shows the layout and photograph of the 16-channel cylindrical virtual sound barrier system used in the experiments.

The performance of the system with $r_c=1.2$ m and $r_e=0.2$ m was measured for 11 sinusoidal noise signals from 200 Hz to 700 Hz with an interval of 50 Hz. As shown in Figure 3.17, the noise reduction level within the quiet zone decreases with the increase of the noise frequency. The noise reduction

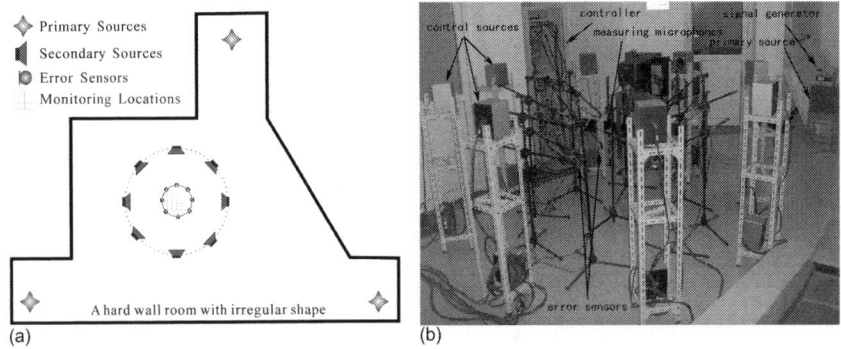

**FIGURE 3.16**
The 16-channel cylindrical virtual sound barrier system used in the experiments: (a) layout, (b) photograph.

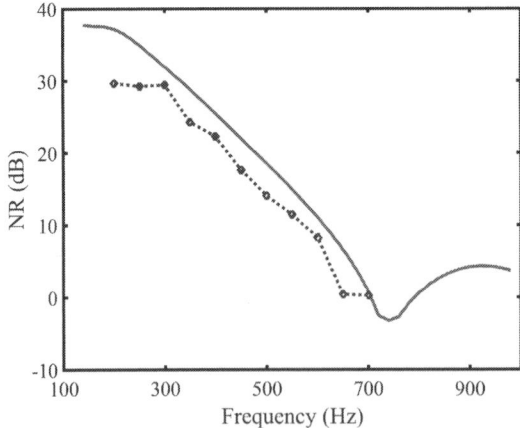

**FIGURE 3.17**
Noise reduction system as a function of frequency of the 16-channel cylindrical virtual sound barrier with $r_c=1.2$ m and $r_e=0.2$ m; the simulation data (solid line) and the experiment data (dotted line marked with diamonds) represent the noise reduction in the cylinder enclosure by the error sensor cylinder.

is greater than 10 dB up to approximate 600 Hz, which has a wavelength of 0.57 m. The 0.4 m diameter of the volume surrounded by the error sensors is about 0.7$\lambda$, which is larger than the $\lambda/10$ size quiet zone for 10 dB attenuation around a single error sensor in a diffuse sound field (Nelson and Elliott, 1992). Being compared with that of active headrest systems in a diffuse sound field, the enlargement of the quiet zone is at the cost of using more secondary sources in the control system.

The control performance of the virtual sound barrier system is affected by the distribution of the error sensors and the secondary sources as well. The control performance with respect to the radius of the error sensor ring $r_e$ (0.1 m, 0.2 m, 0.3 m, 0.4 m, and 0.5 m in the experiments) for the system with $r_c = 1.2$ m at 250 Hz is shown in Figure 3.18. The experiment results agree well with the simulation ones. The control performance deteriorates as the radius of the error sensor ring increases, though it should be sufficiently small (for example, less than 0.2 m) to obtain an effective quiet zone. When the radius of the error sensor ring increases, the distribution of the sound pressure in the enclosed area cannot be reduced uniformly, so the overall noise reduction of the sound pressure in the whole volume decreases.

For comparison, the simulation results with the evaluation cylinder in Section 3.2.4 (the radius of the evaluation cylinder is fixed at 0.2 m) are also drawn in Figure 3.18 with the dotted line. There is an optimum diameter for the error sensor cylinder at about 0.18 m, which is about 0.13$\lambda$ ($\lambda = 1.4$ m is the wavelength of 250 Hz sound). For the error sensor cylinder with a small diameter, the sound field inside the cylinder does reduce; however, as the evaluation cylinder is much larger than the error sensor cylinder, the total

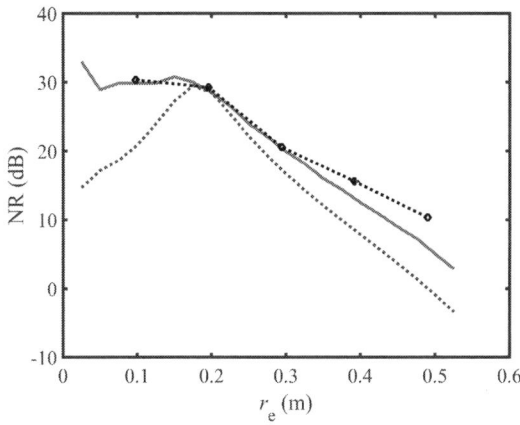

**FIGURE 3.18**
Noise reduction with respect to the radius of the error sensor ring $r_e$ of the 16-channel cylindrical virtual sound barrier system with $r_c = 1.2$ m at $f = 250$ Hz, the simulation data (solid line) and the experiment data (dotted line marked with diamonds) represent the noise reduction in the cylinder enclosure by the error sensor cylinder, while the dotted line represents the noise reduction (simulation) within the evaluation cylinder.

noise reduction in the evaluation cylinder decreases. When the diameter of the error sensor cylinder is larger than half of the wavelength of the sound or the diameter of the evaluation zone, the overall noise reduction inside the evaluation cylinder decreases.

The control performance with respect to $r_c$ is shown in Figure 3.19 for the system with $r_e = 0.38$ m. The radii of the secondary sources are $r_c = 0.65$ m, 0.75 m, 0.85 m, 1.0 m, and 1.2 m in the experiments. The curves in the figure indicate that the control performance improves first, with increasing $r_c$ when $r_c$ is close to $r_e$. After $r_c$ is a certain distance away from the error sensor, the value of the noise reduction begins to decrease. When $r_c$ is close to $r_e$, the distance between an error sensor and the corresponding secondary source is much less than that between the same error sensor and the other secondary sources, so only a small secondary source strength is required to control the sound pressure at the corresponding error sensor. With such a small source strength, the secondary source might only control a small zone, so the sound pressure in the entire cylindrical volume might not be reduced uniformly.

The optimal strengths increase with increasing $r_c$, and this enlarges the "covering zone" of the secondary sources. However, when $r_c$ exceeds a certain value, the secondary sources in the cylinder are far away from the error sensors, and it becomes harder for the secondary sources in this configuration to generate a secondary sound field to match the primary sound that comes from the top and bottom directions, due to the limited height of the secondary source cylinder. Therefore, the noise attenuation performance decreases for this 2.5-dimensional system when the diameter

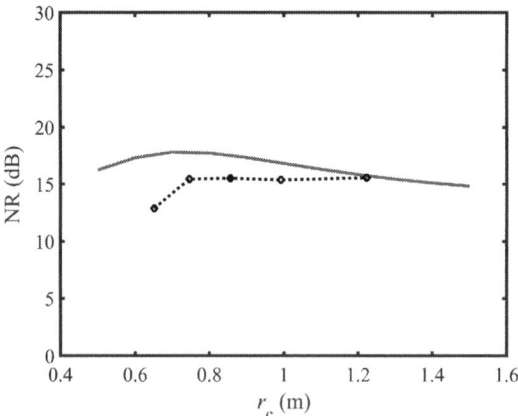

**FIGURE 3.19**

Noise reduction with respect to the $r_c$ of the 16-channel cylindrical virtual sound barrier system with $r_e = 0.38$ m at $f = 250$ Hz; the simulation data (solid line) and the experiment data (dotted line marked with diamonds) represent the noise reduction in the cylinder enclosure by the error sensor cylinder.

of the secondary source cylinder is too large. This is different to the noise attenuation performances in the two- or three-dimensional virtual sound barrier systems.

The differences between the experiment and the simulation results are caused by the differences between the experimental configurations and the numerical model. For example, the sound field in the experiments may not be completely diffuse, so it might differ from that calculated from Equation (3.1). It may also be caused by the simplified processing in numerical simulations where only the direct sound of the secondary sources is taken into account.

### 3.2.6 Remarks

The three-dimensional virtual sound barrier systems have been demonstrated both numerically and experimentally in this section. It is shown that with a 16-channel cylindrical virtual sound barrier system in an ordinary room, the upper-limit frequency can be up to 500 Hz, with an average reduction of more than 10 dB inside a 0.2 m high cylindrical region with a radius of 0.2 m. The noise reduction level within the quiet zone decreases with the increase of noise frequency. The separation distance between two neighboring sensors should be smaller than a half wavelength, and the secondary sources should not be too close to the corresponding error sensors.

As was pointed by Mangiante (1977) in the conclusion of his paper, the main technological difficulties for the implementation of an active noise control (ANC) system are extending the frequency range and making the system automatic. This is the same for virtual sound barrier systems. The physical requirement for the implementation of virtual sound barrier systems in the middle audible frequency range has been illustrated in this section, which shows that it is possible, in principle, to generate a useful size of quiet zone, larger than a human head, up to a middle frequency (for example, 1500 Hz) with a limited number of secondary sources. When the primary sound field consists of a number of plane waves with random phase from many different directions, the average attenuation inside a 0.15 m high cylinder with a diameter of 0.3 m can be up to 11 dB with 24 loudspeakers and 18 microphones. The total increase of average squared sound pressure in a larger cube with dimensions of 2.0 m × 2.0 m × 2.0 m is less than 2 dB.

Future work on the physical aspects of the virtual sound barrier systems includes the optimization of the surrounding acoustical environment, the development of different types of practical secondary sources and error sensors, and the investigation of the mechanisms of the virtual sound barrier systems. Control algorithms and systems also need to be developed, which include the development of decentralized or distributed adaptive ANC algorithms, the development of simple and robust algorithms, studies on the effects of the simplifications on the existing algorithms, and the optimized implementation of the current algorithms on the available hardware.

## 3.3 Performance with a Diffracting Sphere Inside the Quiet Zone

In the applications of a virtual sound barrier system, a person is usually inside the system, so the effects of the human head scattering should be considered. This is modelled by a rigid sphere in this section (Zou and Qiu, 2008).

### 3.3.1 Formulation

Figure 3.20 shows the spherical coordinate system for a rigid sphere with a radius of $a$. The same as that without the sphere, the primary noise field at a location $\mathbf{r} = (r, \theta, \varphi)$ consists of a number of plane waves with random phases from many different directions, which can be expressed as (Lin et al., 2004)

$$p_p(\mathbf{r}) = \frac{1}{\sqrt{N_P}} \sum_{i=1}^{N_P} A_{p,i} e^{-j\varphi_{0i}} \sum_{n=0}^{\infty} (-j)^n (2n+1) \left[ j_n(kr) - a'_n h_n^{(2)}(kr) \right]$$

$$\times \left\{ P_n(\cos\theta_i) P_n(\cos\theta) + 2 \sum_{l=1}^{n} \frac{(n-l)!}{(n+l)!} P_n^l(\cos\theta_i) P_n^l(\cos\theta) \cos[l(\phi - \phi_i)] \right\} \quad (3.11)$$

in the spherical coordinate system, where $j_n(x)$ is the spherical Bessel function of order $n$, $h_n^{(2)}(x)$ is the second kind of spherical Hankel function of order $n$, $a'_n = j'_n(ka)/h_n^{(2)'}(ka)$, $P_n(x)$ is the Legendre function of degree $n$, and $P_n^l(x)$ is the associated Legendre function of degree $n$ and order $l$.

The control sound field generated by $N_s$ secondary monopole sources at point $\mathbf{r}$ with the sphere is

$$p_s(\mathbf{r}) = \sum_{i=1}^{N_s} \frac{k\omega\rho_0 q_{s,i}}{4\pi} \sum_{n=0}^{\infty} (2n+1) \left[ j_n(kr_<) - a'_n h_n^{(2)}(kr_<) \right] h_n^{(2)}(kr_>)$$

$$\times \left\{ P_n(\cos\theta_i) P_n(\cos\theta) + 2 \sum_{l=1}^{n} \frac{(n-l)!}{(n+l)!} P_n^l(\cos\theta_i) P_n^l(\cos\theta) \cos[l(\phi - \phi_i)] \right\} \quad (3.12)$$

where $r_< = \min\{|\mathbf{r}|, |\mathbf{r}_{s,i}|\}$, $r_> = \max\{(|\mathbf{r}|, |\mathbf{r}_{s,i}|)$, $\mathbf{r}_{s,i} = (r_{s,i}, \theta_{s,i}, \varphi_{s,i})$ is the location of the $i$th secondary source. Following the same procedure as that without the sphere in Section 3.2.1, the performance of the virtual sound barrier system with a sphere can be studied, and the difference is that the $i$th element of $\mathbf{Z}_s(\mathbf{r})$ in Equation (3.5) becomes

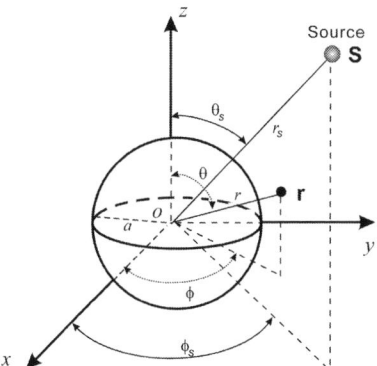

**FIGURE 3.20**
Coordinates of a rigid sphere with a point monopole source.

$$Z_{s,i}(\mathbf{r}) = \frac{k\omega\rho_0}{4\pi} \sum_{n=0}^{\infty} (2n+1)[j_n(kr_<) - a'_n h_n^{(2)}(kr_<)]h_n^{(2)}(kr_>)$$

$$\times \left\{ P_n(\cos\theta_i)P_n(\cos\theta) + 2\sum_{l=1}^{n} \frac{(n-l)!}{(n+l)!} P_n^l(\cos\theta_i)P_n^l(\cos\theta)\cos[l(\phi - \phi_i)] \right\}$$

$$(3.13)$$

With the human head, the sound pressure around the ear positions is important. The ear positions are usually on a horizontal plane when the human head moves and turns around inside the quiet zone, so the points on a horizontal circle including the ear positions on the surface of the head ($r=a$, $z=0$ in Figure 3.20) are chosen as evaluation points, denoted as "ear circle" for short in this section.

### 3.3.2 Simulations and Experiments

The variation of the control performance with and without the diffracting rigid sphere has been analyzed with the 16-channel, cylindrical, virtual sound barrier system described in Section 3.2.5. Figure 3.21 shows the setup and a photo of this system with a rigid sphere, where 16 error sensors are spaced on 2 horizontal planes separated by $h_e$, and 8 error sensors in each plane are evenly spaced in a circle with a radius of $r_e = h_e$. The secondary sources were located in a similar way, surrounding the error sensors in the 2 horizontal planes separated by $h_c$ and the circle radius is $r_c = h_c$. A rigid sphere of radius $a = 0.09$ m was located inside the cylinder surrounded by the error sensors. In the experiments, a hollow iron sphere with a radius of approximate 0.09 m was used. The center of the sphere is at the origin of the coordinates, which is the same as the center of the virtual sound barrier system.

The performance of the virtual sound barrier system with a rigid sphere is affected by the arrangement of the secondary sources and the error sensors, and the trend is similar to that without the rigid sphere. When the diameter of the error sensor cylinder increases, the control performance deteriorates. The control performance increases with the increase in the diameter of the secondary source cylinder when the secondary sources are close to the error sensors, but decreases at larger distances.

The control performance as a function of frequency is shown in Figure 3.22 for the virtual sound barrier system with $r_e = 0.2$ m and $r_c = 1.2$ m. The

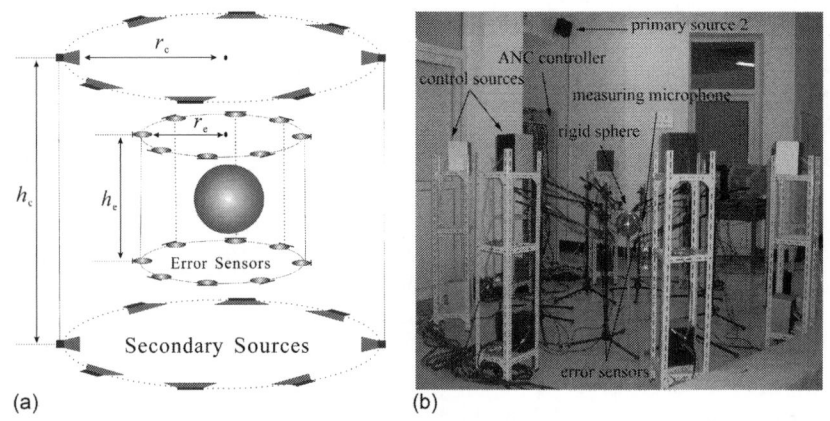

(a)                                               (b)

**FIGURE 3.21**
(a) Configuration of a 16-channel cylindrical virtual sound barrier system with a rigid sphere, (b) photo of the experiment setup.

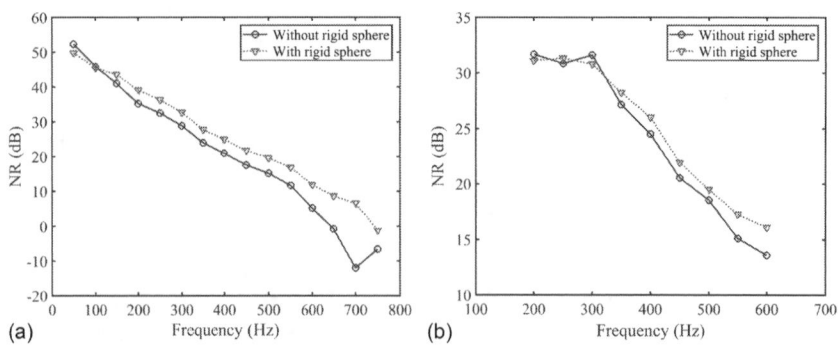

(a)                                               (b)

**FIGURE 3.22**
Control performance as a function of frequency for the virtual sound barrier system with $r_c = 1.2$ m and $r_e = 0.2$ m: (a) simulation, (b) experiment.

changing tendencies of the control performance NR with and without the rigid sphere as a function of frequency are similar, i.e., the NR decreases with the increase of the frequency. The effect of the sphere on performance depends on the frequency. It does not have a significant effect at a frequency below 100 Hz, but has a positive effect (approximate 4 dB more reduction with the rigid sphere) in the frequency range from 100 Hz to 600 Hz. For a practical virtual sound barrier system, in terms of effective noise reduction in the quiet zone, the introduction of the human head is usually beneficial to virtual sound barrier systems.

The distribution of the sound pressure attenuation (simulation) at 603 Hz ($ka=1$) without and with the rigid sphere are shown in Figure 3.23 for the virtual sound barrier system with $r_e=0.2$ m and $r_c=1.2$ m, where (a) and (b) show the sound pressure attenuation in the horizontal annular plane without and with the sphere, while (c) shows their difference. The average pressure attenuation observed at the error sensors is about 33 dB under both conditions. The evaluation points are within the disk with a radius of 0.2 m, excluding the area occupied by the rigid sphere.

Figure 3.23(a) shows that the noise reduction performance without the rigid sphere in the inner boundary is not as good as that close to the outer boundary, and the average sound attenuation on the "ear circle" is 5.4 dB, while the control performance NR is calculated to be 5.2 dB. The average sound attenuation on the circle increases with radius of the circle, and the maximum average attenuation is 18.2 dB at a radius of 0.17 m. The average attenuation on the whole annular plane is 11.9 dB. Figure 3.23(b) shows that the average pressure attenuation on the "ear circle" is 12.2 dB with the rigid sphere, while the control performance NR is 11.9 dB. The average pressure attenuation on the circle at the radius of 0.15 m achieves the maximal value of 16.4 dB while the average pressure attenuation on the whole annular plane is 13.8 dB.

Figure 3.24 shows the simulation results when the radius of the error ring reduces to 0.12 m, where (a) shows that the control performance without

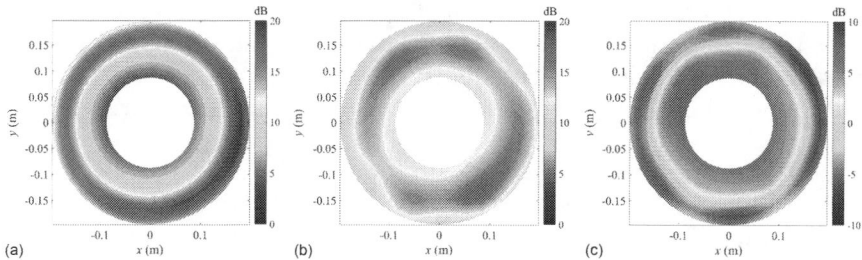

**FIGURE 3.23**
Distribution of the sound pressure attenuation (simulation) in the horizontal annular plane for the virtual sound barrier system with $r_e=0.2$ m, $r_c=1.2$ m, $ka=1$: (a) without the rigid sphere, (b) with the rigid sphere, (c) the difference between the sound pressure attenuation with and without the rigid sphere.

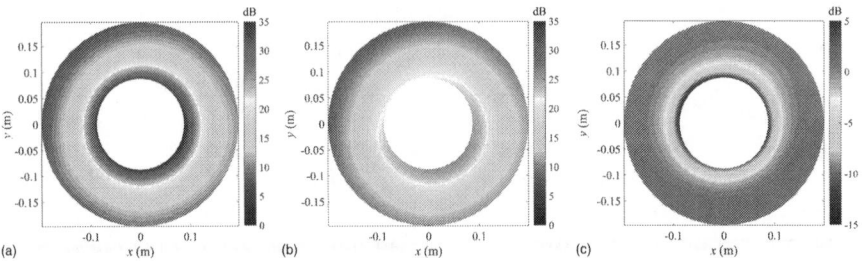

**FIGURE 3.24**
Distribution of the sound pressure attenuation (simulation) in the horizontal annular plane for the virtual sound barrier system with $r_e = 0.12$ m, $r_c = 1.2$ m, $ka = 1$: (a) without the rigid sphere, (b) with the rigid sphere, (c) the difference between the sound pressure attenuation with and without the rigid sphere.

the sphere near the inner boundary is better than that close to the outer boundary. The average pressure attenuation on the "ear circle" is 36.3 dB, while the control performance NR is 34.7 dB. The average attenuation on the circle decreases with the increase of the circle radius. The minimum is 9.4 dB when the circle radius is 0.2 m, which is at the outer boundary of the annular plane. The average pressure attenuation of the whole annular plane is 19.2 dB. Figure 3.24(b) shows that the average pressure attenuation on the "ear circle" is 24.4 dB, while the control performance NR is 23.6 dB. The minimum attenuation is approximately 10.0 dB at the outer boundary of the annular plane, while the average pressure attenuation on the whole annular plane is 17.0 dB.

In Figure 3.23(a), the radius of the error sensor ring is $r_e = 0.2$ m and the noise control near the inner boundary is not as good as that near the outer boundary, while in Figure 3.24(a), $r_e = 0.12$ m and the noise control near the inner boundary is better than that near the outer boundary. Figures 3.23(b) and 3.24(b) show that the variation in the radial direction of the pressure attenuation tends to be smooth and the distribution of the pressure attenuation tends to be uniform with the introduction of the rigid sphere. There are two reasons for this. First, with the rigid sphere, the total control sound field (including the original control sound field and the scattered sound field of the secondary sources) becomes similar to the primary sound field, so the sound attenuation tends to be more uniform in all areas. Second, the normal pressure gradient is zero on the surface of the rigid sphere, so the variation of the sound pressure in the radial direction is small, and this causes the sound attenuation to tend to be uniform in normal direction (Garcia-Bonito, Elliot, and Bonilha, 1997). The control performance NR increases from 5.2 dB to 11.9 dB in Figure 3.23 and decreases from 34.7 dB to 23.6 dB in Figure 3.24 due to the introduction of the rigid sphere, so both the negative and positive effects of a rigid sphere on the performance exist, but the attenuation of the sound pressure tends to be more uniform due to the scattering effect of the rigid sphere.

### 3.3.3 Performance with a Moving Sphere

The rigid sphere (human head) might move in the quiet zone of a practical virtual sound barrier system. The noise reduction performance of a non-adaptive system changes with the rigid sphere movement because the position of the "ear circle" and the transfer functions between the secondary sources and the error sensors (secondary path transfer functions) change with the movement. The rigid sphere is at the center of the virtual sound barrier system when the secondary path transfer functions are measured, so the best performance for the system is achieved when the sphere is at the center of the system. When the rigid sphere moves in the quiet zone, the secondary source strength obtained based on the previous secondary path transfer functions might not be the optimal strength under this new condition.

Figure 3.25 shows the control performance NR with respect to the movements of the rigid sphere at 250 Hz and 500 Hz for the virtual sound barrier with $r_e = 0.2$ m and $r_c = 1.2$ m. The rigid sphere moves in two directions, one is the vertical movement in the $z$ direction and the other is the horizontal

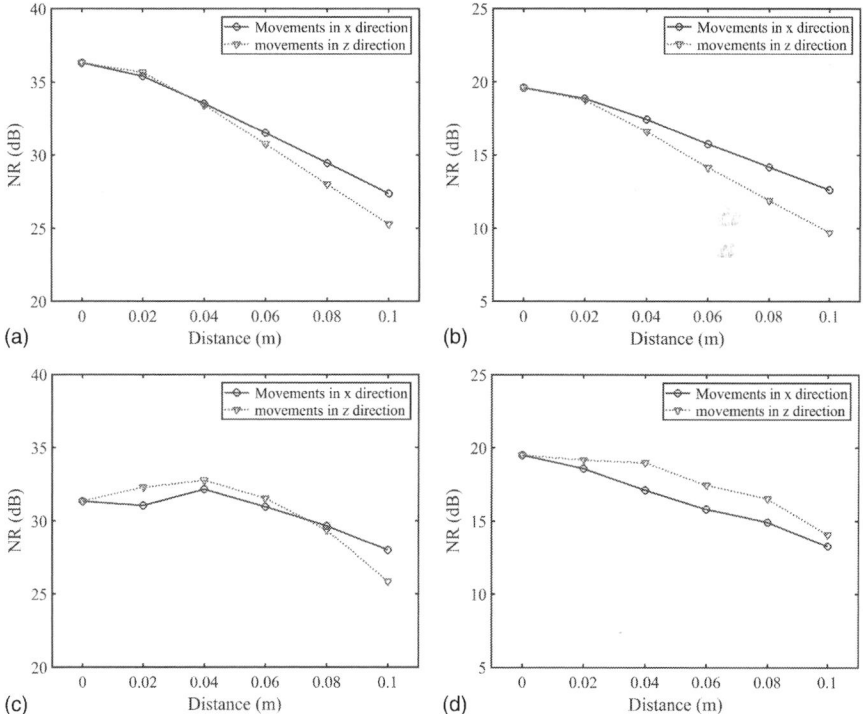

**FIGURE 3.25**

Control performance with respect to the movements of the rigid sphere of the virtual sound barrier system with $r_c = 1.2$ m, $r_e = 0.2$ m: (a) simulation results at 250 Hz, (b) simulation results at 500 Hz, (c) experiment results at 250 Hz, (d) experiment results at 500 Hz.

movement in the $x$–$y$ plane. Figures 3.25(a) and (b) are the simulation results, while (c) and (d) are the experiment results. In the experiments, the moving distance between the center of the sphere and the origin of the coordinates is from 0.02 m to 0.10 m with an interval of 0.02 m. When the movement distance is 0.1 m, the surface of the sphere is close to the boundary of the quiet zone surrounded by the error sensors, and NR is the minimum. Both simulation and experiment results show that the control performance NR decreases when the sphere moves away from the original location. In the experiments, the maximal variation of the noise reduction is 5.5 dB at 250 Hz and 6.2 dB at 500 Hz, but the minimal noise reduction is still greater than 10 dB at 500 Hz, so the virtual sound barrier system performance is quite robust for the sphere movements.

## 3.4 Performance near a Reflective Surface

The effect of the presence of a nearby reflective surface on the active control of sound power radiated by a few point monopole sources was studied by Cunefare and Shepard (1993), where they considered a simple active noise control system consisting of one primary point monopole source and one secondary point monopole source in a line parallel or perpendicular to a rigid or pressure release surface. Their research shows that the surface can significantly affect the performance of active control when the sources to be controlled are within one wavelength of the reflective surface. The orientation of the noise sources and the secondary sources with respect to each other and to the surface influence the control performance significantly. For sources more than one wavelength away from the reflective surface, the use of the simpler free-space analysis yields acceptable results, compared to those using the more complex half-space analysis.

Later, Shepard and Cunefare (1994) further investigated the effect the presence of a rigid surface on the active noise control of an acoustic source with characteristic dimensions comparable to the acoustic wavelength at the frequency of interest. Their numerical results show that the presence of the surface may be neglected when the source is located at a distance of one wavelength or more from the surface; however, the minimum radiated power computed using a free-space active analysis can be significantly greater than that predicted using a half-space analysis for distances less than one wavelength. It has also been observed that, at a particular frequency and certain distances from the surface, the surface significantly affects the system performance when the source center is near one-half or one-quarter wavelengths from the surface, leading to very high predicted error. They pointed out that further exploration of this effect (for an admittedly limited set of noise source distributions) might enhance the effectiveness of active noise control applications greatly through appropriate selection of source configuration and surface separation distance.

Pan and Qiu (2008) investigated the power radiated by an active control system with two monopoles being located near a reflective surface. The radiated power and the optimal input for the controlled monopole were derived analytically and it was found that the power output of such an active noise control system near a reflective surface depended upon the system orientation angle. Based on the mechanism that the reflecting surface can convert a dipole vertical to the surface into a longitudinal quadrupole, it was demonstrated that such a feature may be used to improve the performance of active noise control systems. Although the reflecting surface was modeled as an infinitely large rigid baffle in their paper, there is evidence that the conclusions can be extended to the cases even where the reflecting surface is of finite size.

For all the above studies where the distance between the primary sources and the secondary sources is usually less than a half wavelength of the highest frequency of interest, the main mechanism of the control is to alter the impedance seen by the primary source so that the total power output of primary and secondary sources is minimized. However, for the virtual sound barrier systems discussed in this chapter, the primary sources usually are far away from the control areas, and it is hard to put the secondary sources near the noise source or to couple them effectively, thus the situation is "field control" instead of "source control". The performance of a planar virtual sound barrier in a free field near a reflection surface was investigated by Guo and Pan (1998a), and they found that the reflective surface could affect the quiet zone size and the sound power output of the system significantly.

Section 3.4 discusses the performance of a virtual sound barrier system near a reflective surface in a room where the incident sound comes from many directions. Furthermore, this section introduces a hybrid virtual sound barrier where part of the system is an active noise control system that consists of arrays of loudspeakers and microphones, and the other part of system is a reflective surface (Qiu, Zou, and Rao, 2009).

### 3.4.1 Formulation

Without loss of generality, as shown in Figure 3.26, assume that the rigid reflective surface is at $x = x_w$ and the primary sound field consists of a number of plane waves with random phases from many different directions. The sound pressure at point $\mathbf{r}$ near the rigid plane can be expressed as

$$p_p(\mathbf{r}) = \frac{1}{\sqrt{N_p}} \sum_{i=1}^{N_p} A_{p,i} \left[ e^{-j(k\mathbf{n}_i \cdot \mathbf{r} + \varphi_{0,i})} + e^{-j(k\mathbf{n}_i' \cdot \mathbf{r} + \varphi_{0,i} + 2k_{xi}x_w)} \right] \qquad (3.14)$$

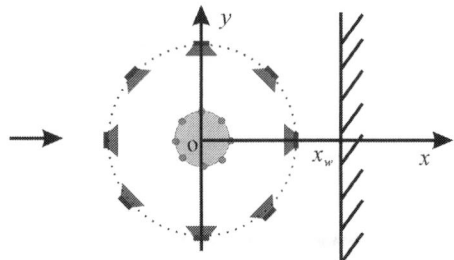

**FIGURE 3.26**
Schematic drawing of a virtual sound barrier system near a reflective surface at $x = x_w$.

where $N_p$ is the number of plane waves, and the $i$th plane wave with magnitude of $A_{p,i}$ comes from a random direction $\mathbf{n}_i$ with a random phase $\varphi_{0,i}$. The magnitude $A_{p,i}$ are taken from uniform distributions of $(0, A_0)$, and $\theta_i$, $\phi_i$, and $\varphi_{0,i}$ are assumed to be uniformly distributed in $(0, -\pi)$, $(0, 2\pi)$, and $(-\pi, \pi)$, respectively. $k = \omega/c_0$ is the wave number, $c_0$ is the speed of sound in the air, $\omega = 2\pi f$ is the angular frequency, and $f$ is the frequency. The propagation direction of the primary sound is $\mathbf{n}_i = (\sin\theta_i\cos\phi_i,\ \sin\theta_i\sin\phi_i,\ \cos\theta_i)$, and the reflective wave has a direction of $\mathbf{n}'_i = (-\sin\theta_i\cos\phi_i,\ \sin\theta_i\sin\phi_i,\ \cos\theta_i)$ and an additional phase of $-2k_{xi}x_w$, where $k_{xi} = k\sin\theta_i\cos\phi_i$. When $N_p$ is sufficiently large, the primary sound field given above is used to approximate a diffuse sound field near a reflective surface (Nelson and Elliott, 1992).

The secondary sound field generated by $N_s$ secondary monopole sources near a reflective surface is

$$p_s(\mathbf{r}) = \sum_{i=1}^{N_s} j\omega\rho_0 q_{s,i}\left(\frac{e^{-jk|\mathbf{r}-\mathbf{r}_{s,i}|}}{4\pi|\mathbf{r}-\mathbf{r}_{s,i}|} + \frac{e^{-jk|\mathbf{r}-\mathbf{r}_{sR,i}|}}{4\pi|\mathbf{r}-\mathbf{r}_{sR,i}|}\right) \tag{3.15}$$

where $\rho_0$ is the air density, $q_{s,i}$ is the complex source strength of the $i$th secondary source located at $\mathbf{r}_{s,i} = (x_{s,i}, y_{s,i}, z_{s,i})$ with an image source located at $\mathbf{r}_{sR,i} = (2x_w - x_{s,i}, y_{s,i}, z_{s,i})$. The total sound pressure at location $\mathbf{r}$ is the superposition of the primary and secondary sounds, which can be expressed in matrix form as

$$p_t(\mathbf{r}) = p_p(\mathbf{r}) + \mathbf{Z}_s(\mathbf{r})\mathbf{q}_s \tag{3.16}$$

with $\mathbf{q}_s = [q_{s,1}, q_{s,2}, \ldots, q_{s,Ns}]^T$, $\mathbf{Z}_s(\mathbf{r}) = [Z_{s,1}(\mathbf{r}), Z_{s,2}(\mathbf{r}), \ldots, Z_{s,Ns}(\mathbf{r})]$, and the $i$th element of $\mathbf{Z}_s(\mathbf{r})$ is

$$Z_{s,i}(\mathbf{r}) = j\omega\rho_0\left(\frac{e^{-jk|\mathbf{r}-\mathbf{r}_{s,i}|}}{4\pi|\mathbf{r}-\mathbf{r}_{s,i}|} + \frac{e^{-jk|\mathbf{r}-\mathbf{r}_{sR,i}|}}{4\pi|\mathbf{r}-\mathbf{r}_{sR,i}|}\right) \tag{3.17}$$

The cost function can be defined in a similar way to that used in Section 3.2.1 to obtain the optimal secondary source strength. The performance of the virtual sound barrier system considered in this section is defined as the ratio of the sum of the squared sound pressure inside a 0.15 m high cylinder with a diameter of 0.3 m without and with control. The cylinder is centered at the origin of the coordinates and the height is on the $z$-axis. The space between two evaluation points is 0.05 m, which provides sufficient precision for sound up to 1500 Hz.

Only two-dimensional simulations are considered, where the primary sound field consists of the plane waves travelling on the $x$–$y$ plane only, and the secondary source array and the error sensor array are in the shape of a ring on the $x$–$y$ plane, centered at the origin, as shown in Figure 3.26. This kind of situation is rarely met in practice as it requires not only that the ceiling and floor are completely absorptive but also that the primary sound sources are in the same horizon as the secondary sources. Although it is known that the performance of the two-dimensional configuration deteriorates dramatically when there is a plane wave from the $z$ direction, this simple model is adopted in this section to focus on the basic and main physics mechanisms. In the simulations, the elevation angle θ is set at 90°, and the control force weighting parameter β in the cost functions is set at 0.01 of the largest eigenvalue of the corresponding matrix **A**.

### 3.4.2 Performance near a Reflective Surface

In the simulations, the perimeter of the evaluation cylinder on the $x$–$y$ plane is about 0.94 m, and half of the acoustical wavelength at 1500 Hz is about 0.12 m, so 8 error sensors and secondary sources are used. The diameters of the circular secondary source array and the error sensor array are 1.0 m and 0.2 m, respectively. Figure 3.27 shows the performance of the virtual sound barrier system when there is no reflective surface (solid line) and, when there is an infinitely large reflective surface 1 m away ($x_w = 1$ m), where the dashed

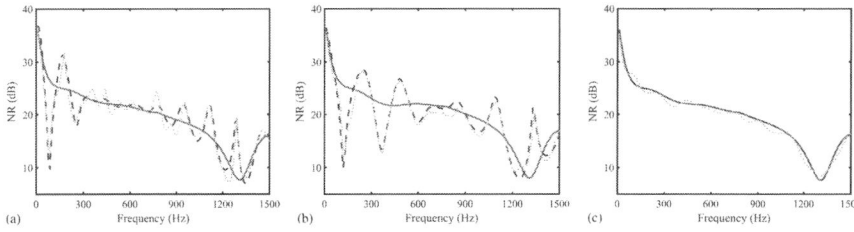

(a)    (b)    (c)

**FIGURE 3.27**
Performance of the virtual sound barrier system when there is no reflective surface (solid line), and when there is an infinitely large reflective surface 1 m away (dashed line, only consider the reflection of the primary sound; dotted line, consider both the reflections of the primary sound and the control sound): (a) normal incidence ф=0°, (b) oblique incidence ф=45°, (c) grazing incidence ф=90°.

line is the one that only takes into account the reflection of the primary sound, while the dotted line considers both the reflections of the primary sound and the control sound. Three different azimuth angles are considered. They are the normal incidence angle with $\phi = 0°$, the oblique incidence angle with $\phi = 45°$, and the grazing incidence angle with $\phi = 90°$.

As can be observed from Figure 3.27, when a virtual sound barrier system is near a reflective surface, its noise reduction performance fluctuates around the curve without a nearby surface. At some frequencies, the surface makes the noise reduction of the system larger while at other frequencies the surface degrades the system performance. The direction of the incidence also affects the performance of the system near the reflective surface. When the incident wave is parallel to the surface (grazing incidence), the effect of the reflective surface is not significant. The fluctuations are mainly caused by the reflection of the primary incidence plane wave because the reflection sound of the secondary sources is smaller than its direct sound when the surface is 1 m away.

For the virtual sound barrier system near an infinitely large reflective surface, with the distance between them changing from 0.5 m to 5.5 m, Figure 3.28 shows the system performance at 250 Hz, 500 Hz, and 1000 Hz for the incidence plane waves from different angles. The effects of the rigid reflective surface on the performance of the virtual sound barrier system vary periodically with the increase of the distance. The period length appears to be equal to the half wavelength for the normal incident sound, becoming larger for oblique incident sounds.

The performance of the virtual sound barrier system near an infinitely large reflective surface with different incidence plane wave angles is shown in Figure 3.29, where the noise reductions at different frequencies are given for when there is no reflective surface and for when the reflective surface is 0.5 m or 1 m away from the center of the system. When there is no reflective surface, the noise reduction of the system is not sensitive to the incident angle in the low-frequency range, even at 1000 Hz. When there is a reflective surface nearby, no matter whether it is 0.5 m away or 1 m away, the variation

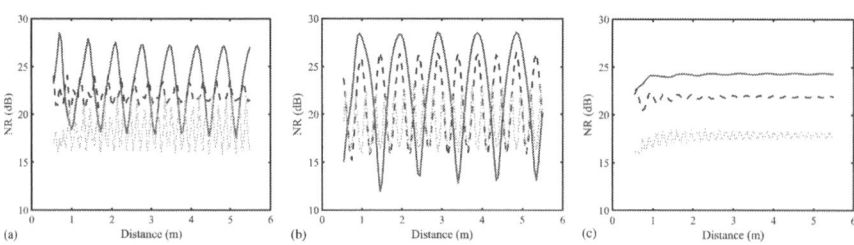

**FIGURE 3.28**
Performance of the virtual sound barrier system near an infinitely large reflective surface with different distances from 0.5 m to 5.5 m (solid line, 250 Hz; dashed line, 500 Hz; dotted line, 1000 Hz): (a) normal incidence $\phi = 0°$, (b) oblique incidence $\phi = 45°$, (c) grazing incidence $\phi = 90°$.

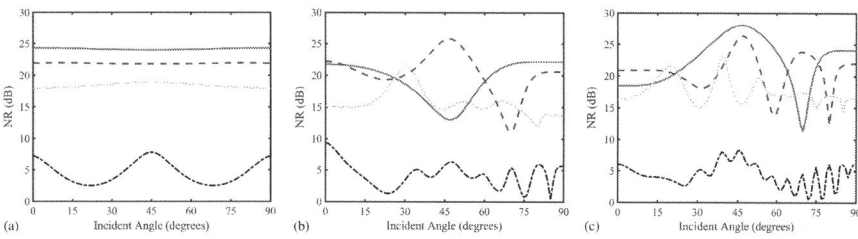

**FIGURE 3.29**
Performance of the virtual sound barrier system near an infinitely large reflective surface with different incidence angles φ from 0° to 90° (solid line, 250 Hz; dashed line, 500 Hz; dotted line, 1000 Hz; dash-dotted line, 2000 Hz): (a) without reflective surface, (b) the reflective surface 0.5 m away, (c) the reflective surface 1.0 m away.

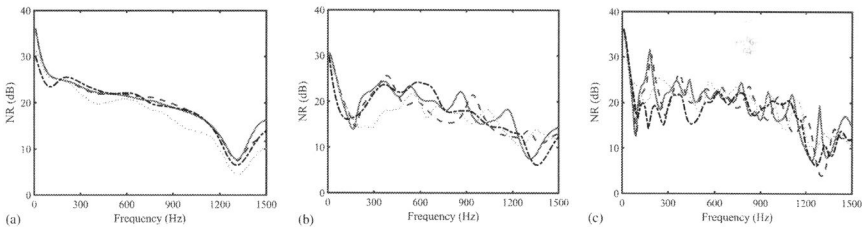

**FIGURE 3.30**
Performance of the virtual sound barrier system near an infinitely large reflective surface with random incidence (solid line, normal incidence φ=0°; dashed line, oblique incidence φ=22.5°; dotted and dash-dotted lines, 2 trials of random incidence): (a) without reflective surface, (b) the reflective surface 0.5 m away, (c) the reflective surface 1.0 m away.

of the virtual sound barrier system performance can be as large as 10 dB even at the low frequency of 250 Hz. It is interesting to note that the variation of the system performance is not significant when the incident angle is less than 15 degrees.

Figure 3.30 shows the performance of the virtual sound barrier system near an infinitely large reflective surface with random incidence. For comparison, the performances of the normal incidence (φ=0°, solid line) and the oblique incidence (φ=22.5°, dashed line) are also given in the figures. The dotted and dash-dotted lines in the figures show the results of two trials of 100 plane waves with random incidence directions. It is clear that although the values are not the same, the trends of the performance of the virtual sound barrier system for random incidence of many plane waves are similar to that of a single plane wave incident from a certain angle.

### 3.4.3 A Hybrid Virtual Sound Barrier near a Surface

Sometimes, a virtual sound barrier system can be very close to a wall to form a hybrid virtual sound barrier with some part being the surface, as shown in Figure 3.31, where Figure 3.31(a) is the original 8-loudspeaker system, with the reflective surface at a distance of $x_w=0.5$ m, Figure 3.31(b) is a 7-loud-speaker system, with the reflective surface 0.4 m away, and Figure 3.31(c) is a 5-loudspeaker system, with the reflective surface 0.3 m away.

The performance of the virtual sound barrier system with the configurations shown in Figure 3.31 under different incidence directions is shown in Figure 3.32. First, the infinitely large surface is assumed to be completely absorptive. The figures show that the 5-loudspeaker system has almost the same performance as that of the 7- and 8-loudspeaker systems when the primary noise is from the direction of $\phi=0°$, while the performance of the 5-loudspeaker system is only a few dBs lower than that with more secondary sources for the grazing incidence with $\phi=90°$.

Figure 3.33 shows the performance of the virtual sound barrier system near an infinitely large reflective surface under different incident directions. By comparing these curves with those in Figure 3.32, it is clear that the performance of the 5-loudspeaker system is generally lower than that with more secondary sources due to the reflection of the surface; however, the performance of the virtual sound barrier system can still be larger than 10 dB below 1000 Hz. At the frequencies below 200 Hz, the performances of the three systems do not show large differences.

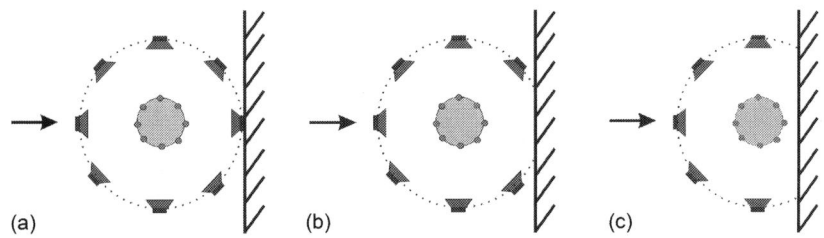

**FIGURE 3.31**
Schematic drawing of a hybrid virtual sound barrier system with some part being a rigid surface: (a) an 8-loudspeaker system, the reflective surface 0.5 m away, (b) a 7-loudspeaker system, the reflective surface 0.4 m away, (c) a 5-loudspeaker system, the reflective surface 0.3 m away.

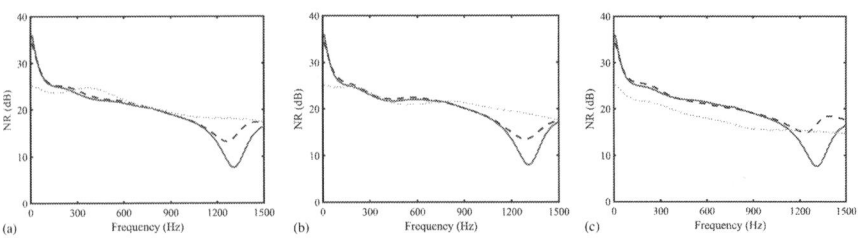

**FIGURE 3.32**
Performance of the virtual sound barrier system near an infinitely large absorptive surface (solid line, the 8-loudspeaker system; dashed line, the 7-loudspeaker system; dotted line, the 5-loudspeaker system): (a) normal incidence $\phi=0°$, (b) oblique incidence $\phi=45°$, (c) grazing incidence $\phi=90°$.

The performance of the virtual sound barrier system near an infinitely large reflective surface under normal incidence $\phi=0°$ for the 5-loudspeaker system with different surface distances is shown in Figure 3.34, where the solid line is for the distance of $x_w=0.4$ m, the dashed line is for the distance of $x_w=0.2$ m, and the dotted line is for the distance of $x_w=0.1$ m. It seems that when the frequency is at about $2nc_0/8x_w$, where $c_0$ is the sound speed and $n$ is an integer from 0, the noise reduction performance has a peak; while for the frequencies at about $(2n+1)c_0/8x_w$, the noise reduction is small. In terms of acoustic wavelength $\lambda$, when the distance $(2x_w)$ is $2n$ times of $\lambda/4$, the maximum noise reduction can be achieved, while when the distance $(2x_w)$ is $(2n+1)$ times of $\lambda/4$, the noise reduction is the minimum.

In summary, when a virtual sound barrier system is near a reflective surface, its noise reduction performance fluctuates around the performance curve without a nearby surface, and values vary periodically with the

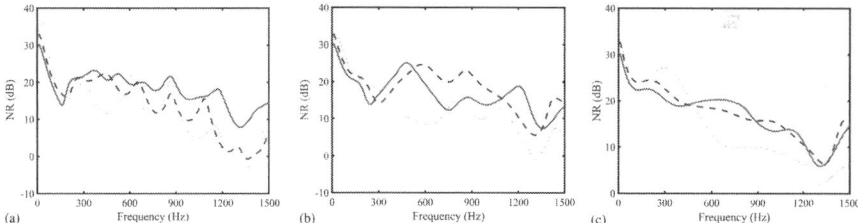

(a)　(b)　(c)

**FIGURE 3.33**
Performance of the virtual sound barrier system near an infinitely large reflective surface (solid line, the 8-loudspeaker system with $x_w=0.5$ m; dashed line, the 7-loudspeaker system with $x_w=0.4$ m; dotted line, the 5-loudspeaker system with $x_w=0.3$ m): (a) normal incidence $\phi=0°$, (b) oblique incidence $\phi=45°$, (c) grazing incidence $\phi=90°$.

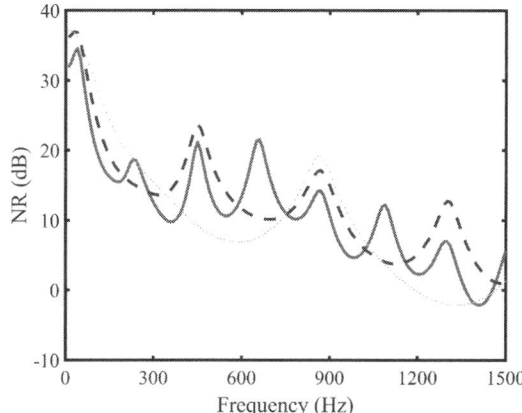

**FIGURE 3.34**
Performance of the 5-loudspeaker virtual sound barrier system near an infinitely large reflective surface under normal incidence $\phi=0°$ (solid line, distance $x_w=0.4$ m; dashed line, distance $x_w=0.2$ m; dotted line, $x_w=0.1$ m).

distance between the surface and the virtual sound barrier system. The fluc-tuation is mainly caused by the change of the primary sound field due to the reflective surface. The reflective surface also makes the noise reduction of the virtual sound barrier become sensitive to the incident angle. For a virtual sound barrier system near an infinitely large absorptive surface, the 5-loud-speaker system has almost the same performance as the system with the 8-loudspeakers; for a virtual sound barrier system near an infinitely large reflective surface, the performance of the 5-loudspeaker system is generally a few dBs lower than that with more secondary sources in the middle fre-quency range due to the reflection of the surface. But at very low frequencies, the performances of the three systems do not show large differences.

## 3.5 Error-Sensing Strategies

To design a virtual sound barrier system, there are a number of parameters to be optimized, such as the number and locations of the secondary sources and error sensors. The optimal strength to drive the secondary sources can be calculated based on a cost function that is related to the objective of con-trol. For applications with time-varying primary and secondary paths, adap-tive algorithms have to be used to adjust the secondary source strength for the minimization of the cost function, which is usually estimated from the error sensor signals. There are a number of different error-sensing strate-gies for predicting the performance of the system based on the error signals. Three cost functions are discussed in this section for virtual sound barrier systems, which include the sum of acoustic potential energy density, the sum of acoustic kinetic energy density, and the sum of the total acoustic energy density at the error sensors (Zou, Qiu, and Lu, 2009).

### 3.5.1 Formulation

In Section 3.2.1, the cost function to be minimized is defined as the sum of acoustic potential energy density

$$J_p = \sum_{i=1}^{N_e} |p_t(\mathbf{r}_{e,i})|^2 + \beta \mathbf{q}_s^H \mathbf{q}_s \tag{3.18}$$

where $N_e$ is the number of error sensors located at $\{\mathbf{r}_{e,i}, i=1, 2, ..., N_e\}$, $p_t(\mathbf{r}_{e,i})$ is the total sound pressure located at $\mathbf{r}_{e,i}$, $\mathbf{q}_s=[q_1, q_2, ..., q_{Ns}]^T$ is the vector of the secondary source strengths, and $\beta$ is a positive real number being used to deter-mine the weighting for the control effort term. The cost function can also be defined as the sum of acoustic kinetic energy density and the sum of the total acoustic energy density at the error sensors (Zou, 2007; Zou, Qiu, and Lu, 2009)

$$J_k = \sum_{i=1}^{N_e} |\mathbf{v}_t(\mathbf{r}_{e,i})|^2 + \beta \mathbf{q}_s^H \mathbf{q}_s \tag{3.19}$$

$$J_e = \sum_{i=1}^{N_e} \left[ \frac{1}{2\rho_0 c_0^2} |p_t(\mathbf{r}_{e,i})|^2 + \frac{\rho_0}{2} |\mathbf{v}_t(\mathbf{r}_{e,i})|^2 \right] + \beta \mathbf{q}_s^H \mathbf{q}_s \tag{3.20}$$

where $c_0$ is the speed of sound in the air, $\rho_0$ is the air density, $\mathbf{v}_t(\mathbf{r})$ is the particle velocity vector. The error sensors are usually located at the boundary of the quiet zone.

The cost functions in Equations (3.18) through (3.20) can be expressed in the quadratic form as

$$J = \mathbf{q}_s^H (\mathbf{A} + \beta I) \mathbf{q}_s + \mathbf{q}_s^H \mathbf{b} + \mathbf{b}^H \mathbf{q}_s + c \tag{3.21}$$

where $\mathbf{A}$ is a matrix related to the transfer functions between the pressure (and/or particle velocity) at the error sensors and the secondary source strengths, $\mathbf{b}$ is a vector related to the transfer functions mentioned above and the primary sound field, and $c$ is a constant related only to the primary sound field.

When the cost function is the acoustic potential energy density shown in Equation (3.18), the total sound field with control is given by

$$p_t(\mathbf{r}) = p_p(\mathbf{r}) + p_s(\mathbf{r}) \tag{3.22}$$

where the primary noise field and the secondary sound field at point $\mathbf{r}$ are the same as in Section 3.2.

$$p_p(\mathbf{r}) = \frac{1}{\sqrt{N_p}} \sum_{i=1}^{N_p} A_{p,i} e^{-j(k\mathbf{n}_i \cdot \mathbf{r} + \varphi_{0,i})} \tag{3.23}$$

$$p_s(\mathbf{r}) = \sum_{i=1}^{N_s} \frac{j\omega\rho_0 q_{s,i}}{4\pi |\mathbf{r} - \mathbf{r}_{s,i}|} e^{-jk|\mathbf{r} - \mathbf{r}_{s,i}|} \tag{3.24}$$

Therefore

$$\mathbf{A} = \mathbf{Z}_s^H \mathbf{Z}_s, \quad \mathbf{b} = \mathbf{Z}_s^H \mathbf{p}_p, \quad c = \mathbf{p}_p^H \mathbf{p}_p, \tag{3.25}$$

$$\mathbf{Z}_s = \begin{bmatrix} Z_{se}(\mathbf{r}_{e,1} | \mathbf{r}_{s,1}) & Z_{se}(\mathbf{r}_{e,1} | \mathbf{r}_{s,2}) & \cdots & Z_{se}(\mathbf{r}_{e,1} | \mathbf{r}_{s,N_s}) \\ Z_{se}(\mathbf{r}_{e,2} | \mathbf{r}_{s,1}) & Z_{se}(\mathbf{r}_{e,2} | \mathbf{r}_{s,2}) & \cdots & Z_{se}(\mathbf{r}_{e,2} | \mathbf{r}_{s,N_s}) \\ \cdots & \cdots & \cdots & \cdots \\ Z_{se}(\mathbf{r}_{e,N_e} | \mathbf{r}_{s,1}) & Z_{se}(\mathbf{r}_{e,N_e} | \mathbf{r}_{s,2}) & \cdots & Z_{se}(\mathbf{r}_{e,N_e} | \mathbf{r}_{s,N_s}) \end{bmatrix} \tag{3.26}$$

$$\mathbf{p}_p = \begin{bmatrix} p_p(\mathbf{r}_{e,1}) & p_p(\mathbf{r}_{e,2}) & \cdots & p_p(\mathbf{r}_{e,N_e}) \end{bmatrix}^{\mathrm{T}} \tag{3.27}$$

where

$$Z_{se}\left(\mathbf{r}_{e,i}\middle|\mathbf{r}_{s,j}\right) = \frac{j\omega\rho_0}{4\pi\left|\mathbf{r}_{e,i}-\mathbf{r}_{s,j}\right|} e^{-jk\left|\mathbf{r}_{e,i}-\mathbf{r}_{s,j}\right|} \tag{3.28}$$

When the cost function is the sum of acoustic kinetic energy density shown in Equation (3.19), the total particle velocity with control is given by

$$\mathbf{v}_t(\mathbf{r}) = \mathbf{v}_p(\mathbf{r}) + \mathbf{v}_s(\mathbf{r}) \tag{3.29}$$

where the particle velocities of the primary and the secondary sound fields at point **r** are

$$\mathbf{v}_p(\mathbf{r}) = \frac{1}{\sqrt{N_p}} \sum_{i=1}^{N_p} \frac{A_{p,i}\mathbf{n}_i}{\rho_0 c_0} e^{-j(k\mathbf{n}_i \cdot \mathbf{r} + \varphi_{0,i})} \tag{3.30}$$

$$\mathbf{v}_s(\mathbf{r}) = \sum_{i=1}^{N_s} \frac{1}{\rho_0 c_0}\left(1 - \frac{j}{k\left|\mathbf{r}-\mathbf{r}_{s,i}\right|}\right) \frac{j\omega\rho_0 q_{s,i}}{4\pi\left|\mathbf{r}-\mathbf{r}_{s,i}\right|} e^{-jk\left|\mathbf{r}-\mathbf{r}_{s,i}\right|} \frac{\mathbf{r}-\mathbf{r}_{s,i}}{\left|\mathbf{r}-\mathbf{r}_{s,i}\right|} \tag{3.31}$$

Therefore

$$\mathbf{A} = \mathbf{Y}_s^{\mathrm{H}}\mathbf{Y}_s, \quad \mathbf{b} = \mathbf{Y}_s^{\mathrm{H}}\mathbf{v}_p, \quad c = \mathbf{v}_p^{\mathrm{H}}\mathbf{v}_p, \tag{3.32}$$

$$\mathbf{Y}_s = \begin{bmatrix} Y_{se}(\mathbf{r}_{e,1}|\mathbf{r}_{s,1}) & Y_{se}(\mathbf{r}_{e,1}|\mathbf{r}_{s,2}) & \cdots & Y_{se}(\mathbf{r}_{e,1}|\mathbf{r}_{s,N_s}) \\ Y_{se}(\mathbf{r}_{e,2}|\mathbf{r}_{s,1}) & Y_{se}(\mathbf{r}_{e,2}|\mathbf{r}_{s,2}) & \cdots & Y_{se}(\mathbf{r}_{e,2}|\mathbf{r}_{s,N_s}) \\ \cdots & \cdots & \cdots & \cdots \\ Y_{se}(\mathbf{r}_{e,N_e}|\mathbf{r}_{s,1}) & Y_{se}(\mathbf{r}_{e,N_e}|\mathbf{r}_{s,2}) & \cdots & Y_{se}(\mathbf{r}_{e,N_e}|\mathbf{r}_{s,N_s}) \end{bmatrix} \tag{3.33}$$

$$\mathbf{v}_p = \begin{bmatrix} v_p(\mathbf{r}_{e,1}) & v_p(\mathbf{r}_{e,2}) & \cdots & v_p(\mathbf{r}_{e,N_e}) \end{bmatrix}^{\mathrm{T}} \tag{3.34}$$

where

$$Y_{se}\left(\mathbf{r}_{e,i}\middle|\mathbf{r}_{s,j}\right) = \frac{1}{\rho_0 c_0}\left(1 - \frac{j}{k\left|\mathbf{r}_{e,i}-\mathbf{r}_{s,j}\right|}\right)\frac{j\omega\rho_0}{4\pi\left|\mathbf{r}_{e,i}-\mathbf{r}_{s,j}\right|} e^{-jk\left|\mathbf{r}_{e,i}-\mathbf{r}_{s,j}\right|}\frac{\mathbf{r}_{e,i}-\mathbf{r}_{s,j}}{\left|\mathbf{r}_{e,i}-\mathbf{r}_{s,j}\right|} \tag{3.35}$$

When the cost function is the total acoustic energy density shown in Equation (3.20), the corresponding parameters are

$$\mathbf{A} = \frac{1}{2\rho_0 c_0^2}\mathbf{Z}_s^H\mathbf{Z}_s + \frac{\rho_0}{2}\mathbf{Y}_s^H\mathbf{Y}_s \tag{3.36}$$

$$\mathbf{b} = \frac{1}{2\rho_0 c_0^2}\mathbf{Z}_s^H\mathbf{p}_p + \frac{\rho_0}{2}\mathbf{Y}_s^H\mathbf{v}_p \tag{3.37}$$

$$c = \frac{1}{2\rho_0 c_0^2}\mathbf{p}_p^H\mathbf{p}_p + \frac{\rho_0}{2}\mathbf{v}_p^H\mathbf{v}_p \tag{3.38}$$

The optimal strength of secondary sources for all above cost functions is given by (Zou, 2007)

$$\mathbf{q}_s = -(\mathbf{A}+\beta I)^{-1}\mathbf{b} \tag{3.39}$$

The performance of the virtual sound barrier systems considered in this section is defined as the ratio of the sum of the squared sound pressure inside a defined volume without and with control

$$\mathrm{NR} = 10\log_{10}\frac{\sum_{i=1}^{N_v}|p_p(\mathbf{r}_{v,i})|^2}{\sum_{i=1}^{N_v}|p_{t,o}(\mathbf{r}_{v,i})|^2} \tag{3.40}$$

where $\mathbf{r}_{v,i}$, $i=1, 2, \ldots, N_v$, are the locations of the evaluation points, and $N_v$ is the number of evaluation points, which is chosen to ensure at least 6 evaluation points per wavelength.

### 3.5.2 Simulations

The same 16-channel cylindrical virtual sound barrier system shown in Figure 3.15 of Section 3.2.5 is used to investigate the error-sensing strategies, where 16 error sensors are spaced on two horizontal planes separated by $h_e$, and 8 error sensors on each plane are evenly spaced in a circle of radius $r_e=h_e$. The secondary sources are located similarly surrounding the error sensors with the two horizontal planes separated by $h_c$ and a circle of radius $r_c=h_c$.

The control performance as a function of frequency with different cost functions is shown in Figure 3.35(a) for the virtual sound barrier system with $r_e=0.2$ m and $r_c=1.2$ m. The three curves show different tendencies. For $J_e$, the control performance deteriorates monotonously as the frequency increases, and the upper-limit frequency is about 680 Hz with the average reduction of more than 10 dB inside the target region. For $J_k$, the curve fluctuates slightly, and the upper-limit frequency is about 570 Hz with the control performance being around 10 dB. With $J_p$, the control performance also decreases with the increasing frequency, and the noise reduction performance is similar to that

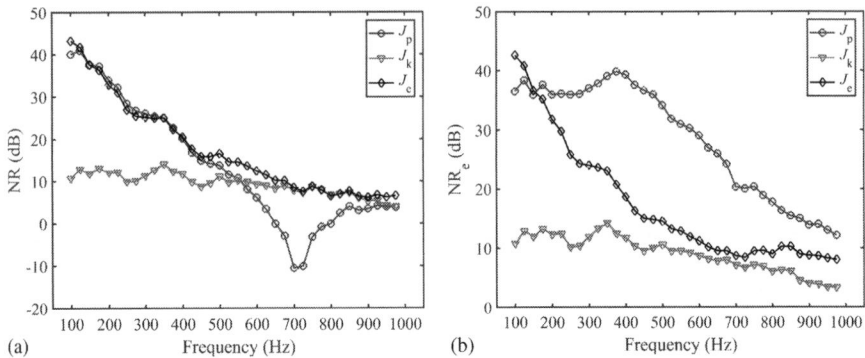

**FIGURE 3.35**
Noise reduction performance as a function of frequency of the noise signal with different cost functions for the virtual sound barrier system with $r_c=1.2$ m and $r_e=0.2$ m: (a) in the whole area, (b) at the error sensor.

with $J_e$ at a frequency lower than 450 Hz; however, its upper-limit frequency of 10 dB quiet zone is only about 550 Hz, and there exists a valley in the curve round 700 Hz where the noise reduction is negative.

The same phenomenon of having a noise reduction valley was also found in the other three-dimensional spherical virtual sound barrier systems when the sum of the squared sound pressure was used as the cost function, and this is related to the interior Dirichlet problem (Epain and Friot, 2007; Kleinman and Roach, 1974). When the frequency is equal to some eigenvalues of an acoustic enclosure, the sound field of the target region is not dependent on the sound pressure on the surface around the target region and the Dirichlet resonance occurs. Thus, the control performance declines dramatically at the frequency around the eigenfrequencies of the interior Dirichlet problem.

Figure 3.35(b) shows the ratio of the sum of the squared sound pressure on the circles without and with control where the error sensors are. For $J_p$, the sound pressure on the boundary is reduced with the sound pressure of the error sensors being controlled in the low frequency; however, the decrease of sound pressure at the error sensors cannot guarantee the effective cancellation of the sound pressure in the whole region, especially at the high frequency. For $J_k$, the particle velocity at the error sensors is minimized, but the sound pressure at the error sensors is not minimized simultaneously due to the item $(1 - j/kr)$ in Equation (3.31). The value of this term approaches 1 at high frequencies, so the performance of $J_k$ is close to $J_p$ at a high frequency.

The cost function $J_e$ is a balance between $J_k$ and $J_p$, as can be observed from the curve of $NR_e$, which is between the other two curves in Figure 3.35(b). The NR in the whole region with $J_e$ is better than the one with $J_p$ because the sound pressure and the particle velocity on the boundary are both controlled. Therefore, the variation of the sound pressure in the target region is small, which makes the sound attenuation uniform. Although the sound attenuation on the boundary is large with the cost function $J_p$, the control in the central area of the target region is weak, so the overall performance with $J_e$ is better than that with $J_p$.

Figure 3.36 shows the sound pressure attenuation distribution in the horizontal annular plane around the center of the target region for the three cost functions at 250 Hz and 550 Hz. In Figure 3.36(a) with $J_p$ as the cost function, the control in the central area is worse than that close to the boundary, and the difference between the maximum and minimum sound attenuation values is about 18 dB. In Figure 3.36(b) with $J_k$, the distribution of the sound pressure attenuation in the target region is smaller but more uniform, and the variation of the sound attenuation is less than 2 dB. In Figure 3.36(c) with $J_e$, the control in the central area is better than that close to the boundary, and the variation of the sound attenuation is about 16 dB. Similar trends can be observed from the simulation results in Figures 3.36(d)–(f) at 550 Hz.

It can be observed from Figure 3.36 that the distribution of sound pressure attenuation is more uniform with the cost function $J_k$. The results from the cost function $J_p$ are the most uneven, while the results for $J_e$ seem to be in the middle between $J_p$ and $J_k$. To describe this noise reduction variation (VR) quantitatively, the sound pressure attenuation uniformity is defined as the root mean square of the sound pressure attenuation inside the sphere surrounded by the error sensors with control as

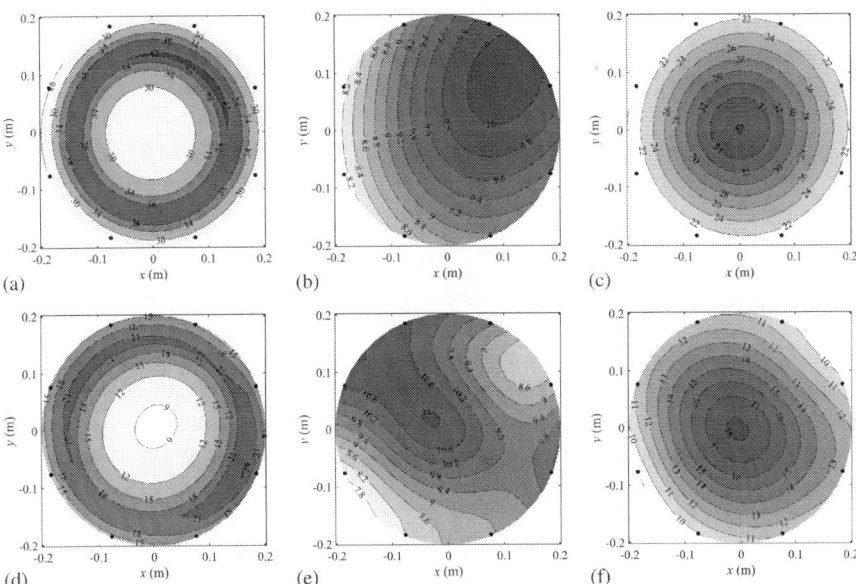

**FIGURE 3.36**
Distribution of the pressure attenuation with different cost functions for the virtual sound barrier system with $r_c = 1.2$ m and $r_e = 0.2$ m: (a, d) $J_p$, (b, e) $J_k$, (c, f) $J_e$, (a–c) are for 250 Hz (d–f) are for 550 Hz, the projections of the error sensors are shown by the point *.

$$VR = \sqrt{\frac{\sum_{i=1}^{N_v}(NR_i - NR)^2}{N_v}} \tag{3.41}$$

where $NR_i = 10\log_{10}(|p_p(\mathbf{r}_i)|^2/|p_t(\mathbf{r}_i)|^2)$ and $\mathbf{r}_i$ is the position of the $i$th point in the region. A smaller VR value means a more uniform sound pressure attenuation in the target region. Figure 3.37 shows the VR with respect to the frequency. It is clear that VR is the largest with the cost function $J_p$, it is the smallest and most uniform with the cost function $J_k$, and that the results with $J_e$ is in the middle between those with $J_p$ and $J_k$.

The numerical simulation results in this section show that good control performance can be achieved by minimizing the sum of the acoustic potential energy density, but the noise reduction distribution is not uniform in the quiet zone. By minimizing the sum of the acoustic kinetic energy density, noise reduction distribution is uniform but has poor control performance. Minimizing the sum of the total energy density is the best strategy in the three cost functions considered, achieving both good control performance and uniform noise reduction distribution.

### 3.5.3 A General Cost Function

A more generalized cost function can be

$$J_g = \sum_{i=1}^{N_e}\left[\alpha_1 \mid p_t(\mathbf{r}_{e,i}) \mid^2 + \alpha_2 \mid \nabla p_t(\mathbf{r}_{e,i}) \mid^2\right] + \beta \mathbf{q}_s^H \mathbf{q}_s \tag{3.42}$$

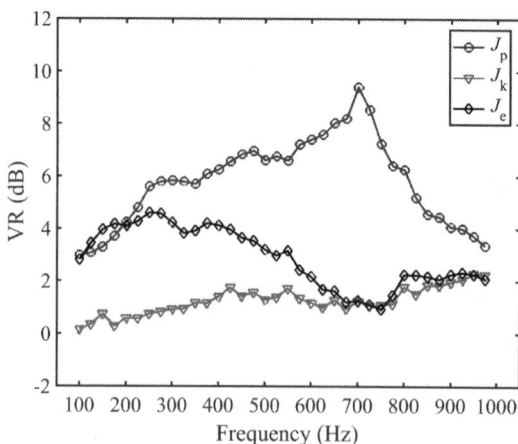

**FIGURE 3.37**
Noise attenuation variance with respect to the frequency with different cost functions for a virtual sound barrier system with $r_c = 1.2$ m and $r_e = 0.2$ m.

where $\alpha_1$ and $\alpha_2$ are the weighting coefficients for the pressure item and the pressure gradient item, respectively. When $\alpha_1=1$ and $\alpha_2=0$, this cost function becomes the one in Equation (3.18) for the sum of the acoustic potential energy density. When $\alpha_1=0$ and $\alpha_2=1$, this cost function is similar to the one in Equation (3.19) for the sum of the acoustic kinetic energy density because the pressure gradient is proportional to the time derivative of the particle velocity. When $\alpha_1=\alpha_2=1/2\rho_0c_0^2$, this cost function is similar to the one in Equation (3.20) for the sum of the total acoustic energy density at the error sensors.

The pressure gradient in the direction of $\mathbf{d}$ can be calculated with the two microphone technique by

$$\nabla_d p(\mathbf{r}) = \frac{p(\mathbf{r}+\mathbf{d})-p(\mathbf{r})}{|\mathbf{d}|} \tag{3.43}$$

where $p(\mathbf{r}+\mathbf{d})$ is the sound pressure at the location $\mathbf{r}+\mathbf{d}$.

The cost function can also be expressed in the quadratic form as

$$J = \mathbf{q}_s^H(\mathbf{A}+\beta I)\mathbf{q}_s + \mathbf{q}_s^H\mathbf{b} + \mathbf{b}^H\mathbf{q}_s + c \tag{3.44}$$

where $\mathbf{A}$ is a matrix related to the transfer functions between the pressure and the pressure gradient at the error sensors and the secondary source strengths, $\mathbf{b}$ is a vector related to the transfer functions and the primary sound field, and $c$ is a constant related only to the primary sound field. They can be expressed by

$$\mathbf{A} = \alpha_1\mathbf{Z}_s^H\mathbf{Z}_s + \alpha_2\mathbf{D}_s^H\mathbf{D}_s \tag{3.45}$$

$$\mathbf{b} = \alpha_1\mathbf{Z}_s^H\mathbf{p}_P + \alpha_2\mathbf{D}_s^H\nabla_d\mathbf{p}_P \tag{3.46}$$

$$c = \alpha_1\mathbf{p}_P^H\mathbf{p}_P + \alpha_2\nabla_d\mathbf{p}_P^H\nabla_d\mathbf{p}_P \tag{3.47}$$

where

$$\mathbf{D}_s = \begin{bmatrix} D_{se}(\mathbf{r}_{e,1}|\mathbf{r}_{s,1}) & D_{se}(\mathbf{r}_{e,1}|\mathbf{r}_{s,2}) & \cdots & D_{se}(\mathbf{r}_{e,1}|\mathbf{r}_{s,N_s}) \\ D_{se}(\mathbf{r}_{e,2}|\mathbf{r}_{s,1}) & D_{se}(\mathbf{r}_{e,2}|\mathbf{r}_{s,2}) & \cdots & D_{se}(\mathbf{r}_{e,2}|\mathbf{r}_{s,N_s}) \\ \cdots & \cdots & \cdots & \cdots \\ D_{se}(\mathbf{r}_{e,N_e}|\mathbf{r}_{s,1}) & D_{se}(\mathbf{r}_{e,N_e}|\mathbf{r}_{s,2}) & \cdots & D_{se}(\mathbf{r}_{e,N_e}|\mathbf{r}_{s,N_s}) \end{bmatrix} \tag{3.48}$$

$$D_{se}(\mathbf{r}_{e,i}|\mathbf{r}_{s,j}) = \frac{j\omega\rho_0}{4\pi|\mathbf{d}|}\left(\frac{e^{-jk|\mathbf{r}_{e,i}+\mathbf{d}-\mathbf{r}_{s,j}|}}{|\mathbf{r}_{e,i}+\mathbf{d}-\mathbf{r}_{s,j}|} - \frac{e^{-jk|\mathbf{r}_{e,i}-\mathbf{r}_{s,j}|}}{|\mathbf{r}_{e,i}-\mathbf{r}_{s,j}|}\right) \tag{3.49}$$

$$\nabla_d\mathbf{p}_P = \begin{bmatrix} \nabla_d p_P(\mathbf{r}_{e,1}) & \nabla_d p_P(\mathbf{r}_{e,2}) & \cdots & \nabla_d p_P(\mathbf{r}_{e,N_e}) \end{bmatrix}^T \tag{3.50}$$

$$\nabla_\mathbf{d} p_\mathrm{p}(\mathbf{r}) = \frac{p_\mathrm{p}(\mathbf{r}+\mathbf{d}) - p_\mathrm{p}(\mathbf{r})}{|\mathbf{d}|} \tag{3.51}$$

The optimal strength of secondary sources for the above cost function is given by

$$\mathbf{q}_\mathrm{s} = -(\mathbf{A} + \beta I)^{-1} \mathbf{b} \tag{3.52}$$

In the applications, the weighting coefficients $\alpha_1$ and $\alpha_2$ can be tuned to approach the best performance.

## 3.6 Virtual Error Sensors

In the preceding sections, the error sensors of the virtual sound barrier systems are placed at the boundary of the target region. Good noise reduction in the target region can be obtained if there are plenty of secondary sources and error sensors according to the Kirchhoff-Helmholtz equation. In practice, the number of error sensors might not be enough at a high frequency, resulting in a deteriorated noise reduction performance in the target region. Under these circumstances, the best locations for the error sensors are in the target region, but this sort of arrangement introduces interference to the human head. In this section, virtual sensing techniques that can sometimes solve the problem are described (Zou and Qiu, 2009; Zou, Qiu, and Li, 2015).

The idea of virtual error sensors comes from the research on active headrest systems. In a diffuse sound field, the shape of the quiet zone (the 10 dB reduction area) has been found to be a sphere centered at the error sensor with a diameter no greater than one-tenth of the wavelength of the sound (Joseph, Elliott, and Nelson, 1994; Garcia-Bonito and Elliott, 1995). To obtain good sound attenuation around a human ear, the error sensor should be located very close to it, which may interfere with the human head. The virtual sensor strategy is introduced to solve the problem, in which a virtual sensor is assumed be installed at the original error sensor location, while a physical sensor is moved away from the ear to avoid the interference to the human head (Garcia-Bonito, Elliott, and Boucher, 1997; Garcia-Bonito and Elliott, 1999; Rafaely, Garcia-Bonito, and Elliott, 1999).

The challenge is to estimate the complex pressure at the virtual sensor location based on the signal obtained at the physical sensor. Based on the high spatial correlation of the low-frequency sound field, early research assumes that the primary sound pressures at the physical sensor location and the virtual sensor location are similar, but this is not true in enclosures due to the spatial rate of sound pressure change caused by higher-order modes (Kestell, Cazzolato, and Hansen, 2000, 2001). It was proposed that the sound pressure at the virtual sensor location should be estimated by either interpolation or extrapolation with the sound pressures of two or three nearby physical sensors (Munn et al., 2003). It was demonstrated that excellent noise reduction at the virtual sensor location could be achieved with this approach, and a moving zone of quiet could be generated in a duct by using their proposed virtual sensing strategy (Petersen et al., 2005, 2007). The drawback of the methods using this interpolation and extrapolation is their complexity due to the multiple physical sensors.

A number of virtual sensing algorithms have been developed for local active noise control systems by using virtual or remote microphone techniques, which estimate the error signal at a location that is remote from the physical error sensor with the physical error signal, the control signal, and knowledge of the system (Moreau et al., 2008). Instead of minimizing the physical error signal, the estimated error signal is minimized with the active noise control system in order to generate a zone of quiet at the virtual location. In virtual sound barrier systems that cannot have sufficient secondary sources and error sensors, the virtual sensors are placed in the target region while the physical sensors are placed at the boundary of the target region. The section analyzes the effect of the virtual sensor locations on the performance of the virtual sound barrier systems.

### 3.6.1 Formulation

Figure 3.38 shows a general block diagram for the virtual and remote sensor algorithms, where the reference signal $x(n)$ can be obtained directly from the noise source. $\mathbf{P}_T(z)$ represents the transfer function of the true physical structural and/or acoustical path, while $\mathbf{P}_V(z)$ represents that of the virtual primary path. $\mathbf{S}_T(z)$ represents the transfer function of the secondary path corresponding to the response of the physical error sensor to the control output, while $\mathbf{S}_V(z)$ represents that of the virtual secondary path. Both the virtual

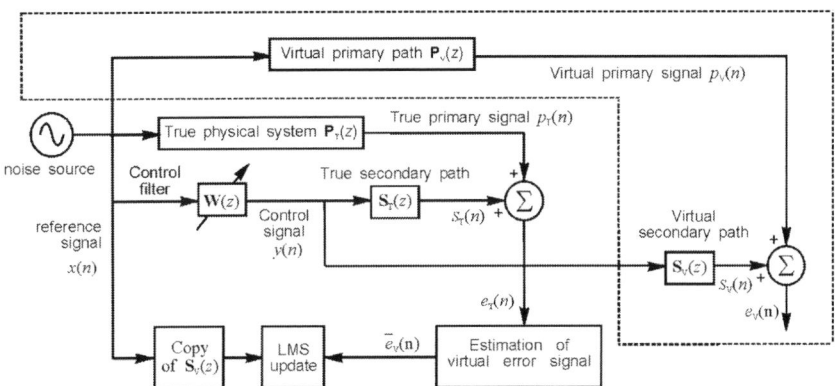

**FIGURE 3.38**
Block diagram for the virtual and remote microphone algorithms.

primary and the virtual secondary paths are determined according to the control objective of the system in the design stage, and the corresponding virtual primary signal $p_V(n)$ and virtual error signal $e_V(n)$ cannot be measured directly in the control stage. The challenge of the virtual sensor or remote sensor algorithms is to obtain an estimation of the virtual error signal $\bar{e}_V(n)$ from the measured true error signal $e_T(n)$.

The virtual sensor algorithms assume that the primary signal at the virtual sensor location is equal to that at the physical location, i.e., $p_V(n)=p_T(n)$. By using

$$e_V(n) = p_V(n) + \mathbf{S}_V(z)y(n) \tag{3.53}$$

$$e_T(n) = p_T(n) + \mathbf{S}_T(z)y(n) \tag{3.54}$$

where $y(n)$ is the control output, it yields

$$e_V(n) = e_T(n) + [\mathbf{S}_V(z) - \mathbf{S}_T(z)]y(n) \tag{3.55}$$

When implemented, the transfer functions of the real and virtual secondary paths are both obtained in advance, so that the virtual error signal $\bar{e}_V(n)$ can be estimated from the measured true error signal $e_T(n)$ using Equation (3.55), which is further used by the filtered-x LMS algorithm for control filter update. Because the transfer function of the virtual secondary path and the estimated virtual error signal are used to adjust the control filter coefficients, the controller minimizes the virtual error signal instead of the real error signal measured by the physical sensor.

The remote sensor algorithms do not assume that the primary signal at the virtual sensor location is equal to that at the physical location, but require a transfer function $\mathbf{T}_p(z)$ from the real primary signal to the virtual primary signal, which can be expressed as (Roure and Albarrazin, 1999),

$$p_V(n) = \mathbf{T}_P(z)p_T(n) \tag{3.56}$$

Substituting Equation (3.56) into Equation (3.53) and using Equation (3.54), gives

$$e_V(n) = \mathbf{T}_P(z)[e_T(n) - \mathbf{S}_T(z)y(n)] + \mathbf{S}_V(z)y(n) \tag{3.57}$$

so

$$e_V(n) = \mathbf{T}_P(z)e_T(n) + [\mathbf{S}_V(z) - \mathbf{T}_P(z)\mathbf{S}_T(z)]y(n) \tag{3.58}$$

When implemented, in addition to knowing the transfer functions of the real and virtual secondary paths, the transfer function from the real primary signal to the virtual primary signal is also required in advance, so that the virtual error signal $\bar{e}_V(n)$ can be estimated from the measured true error signal

$e_T(n)$ with Equation (3.58). This method has been used for optimum control in an active system for local sound control in a spatially random primary field (Elliott and Cheer, 2015).

### 3.6.2 Simulations

Figure 3.39 shows the setup of the 16-channel cylindrical virtual sound barrier system discussed in previous sections, where 16 physical sensors are spaced on two horizontal planes separated by $h_a$, and 8 physical sensors in each plane are evenly spaced in a circle with a radius of $r_a = h_a$. The secondary sources are located similarly, surrounding the physical sensors, with the two horizontal planes separated by $h_c$; the circle radius is $r_c = h_c$. The virtual sensors are located similarly inside the cylinder of the physical sensors, with the two horizontal planes separated by $h_v$; the circle radius is $r_v = h_v$, where $r_v < r_a < r_c$.

The primary noise field at the location $\mathbf{r}$ consists of a number of plane waves with random phases from many different directions. The secondary sources are point monopoles and only their direct sound is taken into account for simplicity. The virtual sound barrier system consists of $N_s$ secondary sources, $N_a$ physical sensors, and $N_v$ virtual sensors ($N_a = N_v$). The sound pressure vector at the virtual sensor locations with control can be calculated by

$$\mathbf{e}_v(\mathbf{r}) = \mathbf{p}_{p,v}(\mathbf{r}) + \mathbf{Z}_{s,v}(\mathbf{r})\mathbf{q}_s \tag{3.59}$$

where $\mathbf{p}_{p,v}(\mathbf{r})$ is the sound pressure vector at the virtual sensor locations due to the primary noise field, $\mathbf{Z}_{sv}(\mathbf{r})$ is an $N_v \times N_s$ matrix of transfer functions from

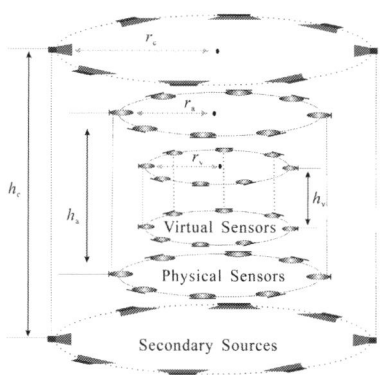

**FIGURE 3.39**
A 16-channel virtual sound barrier system with virtual sensors.

the secondary sources to the virtual sensor locations, and $\mathbf{q}_s$ is the strength of the secondary sources. In the virtual sensor strategy, the sound pressure vector at the virtual sensor locations is assumed to be equal to that at the physical sensor locations, so the above equation becomes,

$$e_v(\mathbf{r}) \approx \mathbf{p}_{p,a}(\mathbf{r}) + \mathbf{Z}_{sv}(\mathbf{r})\mathbf{q}_s \tag{3.60}$$

The sum of the squared sound pressures at virtual sensor locations is selected as the cost function

$$J = e_v^H e_v + \beta \mathbf{q}_s^H \mathbf{q}_s \tag{3.61}$$

and the optimal vector of the secondary source strength is given by

$$\mathbf{q}_s = -(\mathbf{A} + \beta \mathbf{I})^{-1}\mathbf{b} \tag{3.62}$$

where

$$\mathbf{A} = \mathbf{Z}_{sv}^H \mathbf{Z}_{sv}, \quad \mathbf{b} = \mathbf{Z}_{sv}^H \mathbf{p}_p \tag{3.63}$$

The performance of the virtual sound barrier system is defined as the ratio of the sum of the squared sound pressure inside the target region surrounded by the original error sensors without and with control as

$$NR = 10\log_{10} \frac{\sum_{i=1}^{N_v} |p_p(\mathbf{r}_{v,i})|^2}{\sum_{i=1}^{N_v} |p_{t,o}(\mathbf{r}_{v,i})|^2} \tag{3.64}$$

where $\mathbf{r}_{v,i}$, $i=1, 2, ..., N_v$, are the locations of the evaluation points, and $N_v$ is the number of evaluation points, which is chosen to ensure at least 6 evaluation points per wavelength. The radius of the target region is set as 0.2 m.

For the virtual sound barrier systems without using virtual sensing, the error sensors can be located at the best locations in the target region to achieve the maximal noise reduction. Figure 3.40 shows the control performance NR with respect to the physical error sensor cylinder's radius ($r_a$) at three frequencies for the system with $r_c=1.2$ m. The best locations of the error sensors are at the boundary of the target region ($r_a=0.2$ m) at 100 Hz, and move toward the center of the target region at higher frequencies. The best error sensor locations for 550 Hz is $r_a=0.128$ m where the NR is 17.7 dB, while the NR is only 10.8 dB with the error sensors at $r_a=0.2$ m. For this cylindrical virtual sound barrier system, the performance is usually good at low frequencies and poor at high frequencies, so it is desirable to improve the noise reduction performance in the high-frequency range. Therefore, the optimal location of the error sensors could be chosen at $r_a = 0.128$ m for the frequency range below 550 Hz.

The optimal locations for the virtual sensors should be selected at the optimal locations for the physical error sensors, and the practical physical sensors can only be placed at the boundary of the target region. The performance NR as a function of frequency under four circumstances is illustrated in Figure 3.41, where the circumstances include installing physical error sensors at the boundary of the target region ($r_a = 0.2$ m), installing the physical error sensors at the optimal locations for the physical error sensors

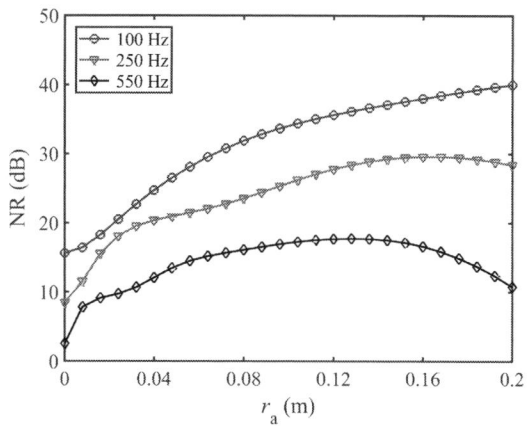

**FIGURE 3.40**
Control performance with respect to the radius of the physical error sensor cylinder for a virtual sound barrier system with $r_c = 1.2$ m.

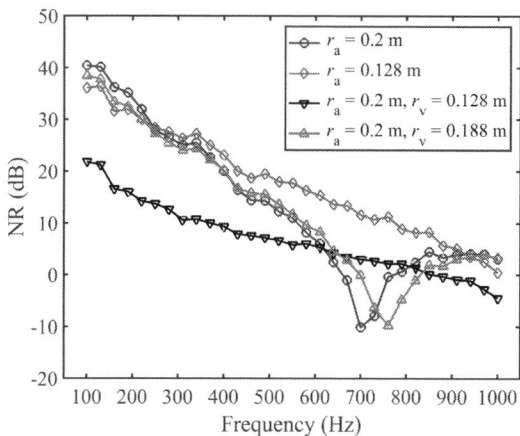

**FIGURE 3.41**
Control performance as a function of frequency for the virtual sound barrier system with $r_c = 1.2$ m.

($r_a$=0.128 m), using the virtual sensors at the optimal locations for the physical error sensors and installing the physical sensors at the boundary of the target region ($r_v$=0.128 m, $r_a$=0.2 m), using the virtual sensors at the optimal locations for the virtual error sensors and installing physical sensors at the boundary of the target region ($r_v$=0.188 m, $r_a$=0.2 m).

The upper-limit frequency of the 10 dB quiet zone, which means 10 dB average sound pressure reduction acquired inside the target region, is about 770 Hz when the physical error sensors are at the optimal locations. This is higher than that of 560 Hz when the physical error sensors are at the boundary of the target region. However, the control performance with the virtual sensors at the optimal locations for the physical error sensors is much worse than that under the two circumstances using only physical error sensors, and the upper-limit frequency of 10 dB quiet zone is only approximately 370 Hz. This is because the sound pressures at the virtual sensor locations cannot be reduced to zero due to the estimation error caused by the assumption that the primary sound pressure at the physical sensor locations is the same as that at the virtual sensor locations.

When the distance between the virtual sensors and the physical sensors increases, the estimation error becomes larger due to the decrease of the sound field spatial correlation. Therefore, the sound pressure attenuation at the virtual sensor locations decreases, resulting in the decrease of the sound pressure reduction inside the entire target region. When the virtual sensor locations are moved away from the boundary of the target region to the physical optimal locations, on the one hand, the performance of the system increases because they are closer to the optimal locations, but, on the other hand, the performance of the system decreases due to the increasing distances between the physical sensors and the virtual sensors. So there exists an optimal location between the boundary of the target region and the physical optimal locations.

The control performance NR with respect to the radius $r_v$ of the virtual sensor cylinder at three frequencies is shown in Figure 3.42 for the system with $r_c$=1.2 m and $r_a$=0.2 m. The optimal location for the virtual sensors at 550 Hz is at $r_v$=0.188 m, which is close to the boundary of the target region. The NR is 11.7 dB, about 1 dB larger than that with the virtual error sensors at the boundary of the target region. Comparing the NR obtained with the virtual sensors at the optimal locations for the physical sensors and the physical sensors at the boundary of the target region ($r_v$=0.188 m, $r_a$=0.2 m) with the NR obtained with the physical error sensors at the boundary of the target region ($r_a$=0.2 m) in Figure 3.41, the upper-limit frequency of 10 dB quiet zone is about 570 Hz in the former case, only 10 Hz higher than that using the physical error sensors. Therefore, when the distance between the secondary sources and the physical sensors is large, the performance improvement achieved with the virtual sensors is not significant.

In practice, a compact virtual sound barrier system is desired, which requires small distances between the secondary sources and the physical sensors. However, as shown in Section 3.2, the control performance decreases dramatically when the secondary sources are very close to the error sensors. The reason is that the secondary source strengths are small under this circumstance and only a small "covering zone" is created with them. The virtual sensor locations can be farther away from the secondary sources than the physical sensors, so the virtual sensing strategy can be used to solve this problem.

The optimal locations of the physical error sensors are calculated first. Figure 3.43 shows the control performance NR with respect to the radius $r_a$ of the physical error sensor cylinder at three frequencies for the system

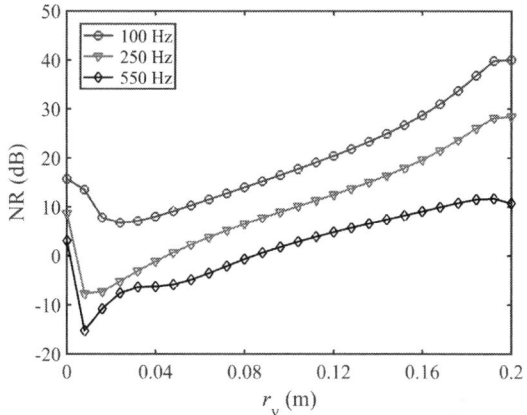

**FIGURE 3.42**
Control performance with respect to the radius of the virtual error sensor cylinder for the virtual sound barrier system with $r_c = 1.2$ m and $r_a = 0.2$ m.

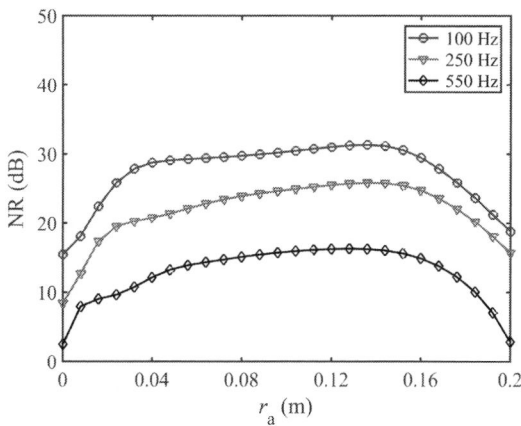

**FIGURE 3.43**
Control performance with respect to the radius of the physical error sensor cylinder for a virtual sound barrier system with $r_c = 0.25$ m.

with $r_c=0.25$ m. Being different from the case with $r_c=1.2$ m, the optimal locations of the physical error sensors are all inside the target region, but not at the boundary of the target region due to the smaller distances between the secondary sources and the boundary of the target region. In the figure, the control performance improves rapidly with the decreasing $r_a$ at first, and then becomes steady when $r_a$ is less than some particular value. Therefore, in this circumstance, the dominant factor for improving the performance is increasing the "covering zone" of the secondary sources, which is different from that for the system with $r_c=1.2$ m. The best arrangement of the physical error sensors is at $r_a=0.128$ m at 550 Hz, with NR of 16.3 dB. It should be noted that NR is only 2.9 dB at 550 Hz for $r_a=0.2$ m, where the error sensors are located at the boundary of the target region.

When the distance between the secondary sources and the physical sensors is large, the system performance with the virtual sensor system is affected by the distance between the virtual sensor locations and the optimal locations of the virtual sensors, which can have a better performance with a smaller value. The system performance is also affected by the distances between the virtual sensor locations and the physical sensor locations, which has a worse performance for a larger value due to the larger evaluated error. When the distances between the secondary sources and the physical sensors is small, the performance is affected by the "covering zone" of the secondary sources.

The NR with respect to the radius $r_v$ of the virtual error sensor cylinder at three frequencies is shown in Figure 3.44 for the system with $r_c=0.25$ m and $r_a=0.2$ m. The NR is only 2.9 dB for $r_v=0.2$ m (equivalent to that without the virtual sensing), and it increases significantly when $r_v$ decreases, and achieving the maximum of 11.3 dB when $r_v=0.178$ m (the optimal locations for the virtual sensors). At this distance, although all the factors mentioned above influence the performance, the third factor is the dominant one, but as $r_v$ decreases further, its becomes less important, and the second one becomes more significant. As a result, the control performance of the system decreases. Using virtual sensing achieves a better performance when $r_v$ is between 0.11 m to 0.2 m, and the average noise reduction inside the target region exceeds 10 dB when $r_v$ is between 0.166 m to 0.186 m.

The system performance NR as a function of frequency under four circumstances is illustrated in Figure 3.45, where the circumstances include installing the error sensors at the boundary of the target region ($r_a=0.2$ m), installing the error sensors at the optimal locations for the physical sensors ($r_a=0.128$ m), using the virtual sensors at the optimal locations for the physical sensors and installing the physical sensors at the boundary of the target region ($r_v=0.128$ m, $r_a=0.2$ m), using the virtual sensors at the optimal locations for the virtual sensors and installing the physical sensors at the boundary of the target region ($r_v=0.178$ m, $r_a=0.2$ m). The performance is the best with the physical error sensors at the optimal locations and the upper-limit frequency of 10 dB quiet zone is about 770 Hz.

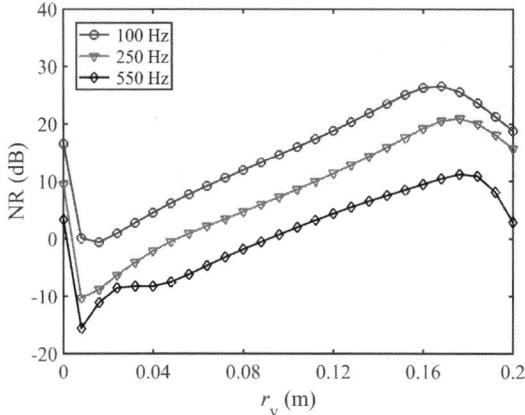

**FIGURE 3.44**
Control performance with respect to the radius of the virtual error sensor cylinder for the virtual sound barrier system with $r_c = 0.25$ m and $r_a = 0.2$ m.

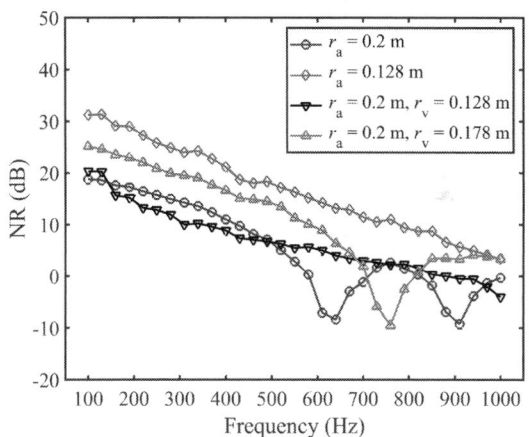

**FIGURE 3.45**
Control performance as a function of frequency for the virtual sound barrier system with $r_c = 0.25$ m.

The performance is poor with the physical error sensors at the boundary of the target region, which is 10 dB lower than that with the physical error sensors at the optimal locations for the physical sensors, and where the upper-limit frequency of 10 dB quiet zone is about 420 Hz. The worst performance is that with the virtual sensors at the optimal locations for the physical sensors, and where the upper-limit frequency is only 350 Hz. A better performance is achieved when the virtual sensors are at the optimal locations for the virtual sensors, and NR is 5 dB higher than that with the physical error sensors at the boundary of the target region below 550 Hz. With this arrangement, the upper-limit frequency of 10 dB quiet zone is about 580 Hz. Therefore, when the distances between the secondary sources and the physical sensors are small, the system performance can be improved significantly with the virtual sensors at their optimal locations. The performance is better than that with the physical error sensors at the boundary of the target region.

### 3.6.3 Remarks

In summary, for the virtual sound barrier systems without sufficient secondary sources and error sensors, the noise reduction in the target region is maximal when the physical error sensors of the system are located at the particular optimal locations in the target region. However, this sort of arrangement results in interference by the error sensors for the human head. This situation can be improved by using the virtual sensing techniques, in which the virtual sensors are "placed" in the target region while the physical sensors are installed at the boundary of the target region.

The best locations for the virtual sensors should be the optimal locations for the physical error sensors if the primary sound pressure at the virtual sensor locations is the same as that at the physical sensor locations. Unfortunately, this assumption does not always hold, especially for the high-frequency sound, so there exist optimal locations for the virtual sensors. The performance of the virtual sound barrier systems with the virtual sensors in their optimal locations is better than the performance when they are placed in the optimal locations for the physical error sensors. When the distance between the virtual sensors and the physical sensors is large, the estimation error is large, so the noise reduction at the virtual sensors is lower, resulting in the noise reduction decrease in the target region. Therefore, the optimal locations for the virtual sensors are usually close to the boundary.

When the distance between the secondary sources and the physical sensors is large, the optimal locations for the virtual sensors are close to the boundary of the target region, so the performance of the virtual sound barrier systems with the virtual sensors placed in their optimal locations is hardly improved, compared to that of placing the error sensors at the boundary of the target region without using the virtual sensors. However, when the distance between the secondary sources and the physical sensors is small, the system performance can be improved significantly. For the 16-channel cylindrical virtual sound barrier system investigated in this section, the noise reduction in the target region (0.2 m height and 0.2 m radius) increases more than 5 dB in the frequency range of less than 550 Hz, and the upper-limit frequency of creating the 10 dB quiet zone increases from 420 Hz to 580 Hz when using the virtual sensors. Therefore, the introduction of the virtual sensors is beneficial for developing compact virtual sound barrier systems.

# 4

## Applications

## 4.1 Noise Radiation Control from Power Transformers in a Hemi-Closed Space

The power transformer noise is dominated by the fundamental frequency and its harmonics, and most of the noise components to be controlled are in the low-frequency range. Sound barriers are usually built around outdoor transformers to block the transformer noise. However, the noise control performance of the sound barriers is poor in the low-frequency range due to the sound diffraction at the edge of the barriers (see Section 1.2). Active control is an alternative solution. Early experiments before and around the 1980s used adjustable phase shifters and power amplifiers to tune the system manually (Ross, 1978; Hesselmann, 1978), while the later experiments normally adopted adaptive control in the systems. The effectiveness of the active control of transformer noise has been validated by theoretical studies, simulations, and field experiments, but it still lacks a mature product with reasonable cost, acceptable performance, and long-term running capability (Qiu, Zhang, and Tao, 2012; Zou, Tao, and Qiu, 2017).

Some transformers are located inside enclosed rooms, but have problems with poor ventilation. In practice, for convenient maintenance and ventilation, some transformers are installed in a hemi-closed space with one or two side openings. In these situations, the transformer noise is mainly transmitted out through the openings (Tao, Wang, and Qiu, 2015; Tao et al., 2016). A planar virtual sound barrier system formed by a loudspeaker array and a microphone array can be arranged at the openings of the enclosure to block the noise radiation from the enclosure as shown in Figure 4.1. The whole structure is a hybrid active and passive noise control system, where the noise radiation to outside is constrained to the opening by the walls and ceiling with large transmission loss, and the virtual sound barrier is installed only on the opening to reduce the sound radiation through it.

Figure 4.1(a) shows a schematic drawing of a planar virtual sound barrier system for this application, while Figure 4.1(b) shows a photo of a 44-channel system installed for two 110 kV transformers in a hemi-closed room in China (Tao, Qiu, and Pan, 2015). The dimensions of the room opening for ventilation and maintenance are approximately 8.3 m×2.8 m. In the virtual sound barrier system, the

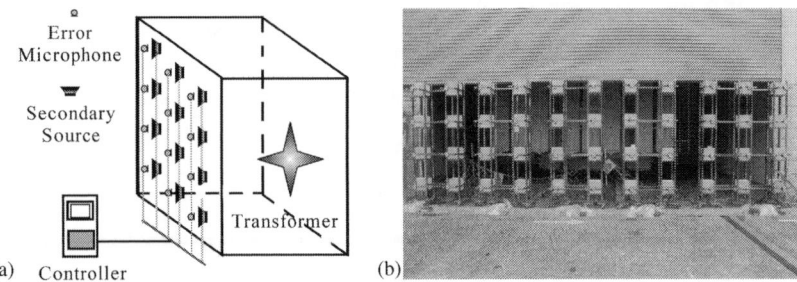

**FIGURE 4.1**

A planar virtual sound barrier system for controlling noise radiation from a transformer inside a hemi-closed space: (a) a schematic drawing, (b) a photo of a 44-channel system installed for 2 110 kV transformers in an enclosure in Guilin, China.

interval between the control loudspeakers is approximately 0.75 m, so a total of 44 control loudspeakers are placed in a lattice structure in 4 rows and 11 columns. The control loudspeaker array is placed 0.2 m outside the opening for safety reasons, and the interval between a control loudspeaker and its collocated error microphone is 0.2 m. The control loudspeakers and error microphones are connected to a 44-channel adaptive controller, which uses a decentralized multiple-channel filtered-x LMS algorithm. The reference signal is obtained by using the self-generated 100 Hz sinusoid signals, which is acceptable in this application because the frequency shift of the transformer noise peaks is within 0.5% in the low-frequency range (Zhang, Tao, and Qiu, 2012).

The noise reduction performance is evaluated on a horizontal plane ($x$–$y$ plane) at a height of 1.2 m. The projection of the error microphone at the far left column is chosen as the coordinate origin, 55 points with a 0.8 m interval in the $x$-axis (a horizontal line parallel to the opening surface) and a 1.0 m interval in the $y$-axis (a horizontal perpendicular to the opening surface) are chosen as evaluation points. Figure 4.2 illustrates the sound pressure level measured at these evaluation points. The maximum sound pressure level is approximately 63.3 dB and 58.6 dB at 100 Hz without and with the planar virtual sound barrier. The average sound pressure level is reduced approximately by 18 dB. Better noise reduction can be obtained if the system is optimized and more powerful adaptive control systems are used.

## 4.2 Sound Transmission Control through an Open Window into a Room

There has been a lot of research for developing a special window that has both good noise reduction performance and sufficient ventilation for providing fresh air. Early in the 1970s, Ford and Kerry (1973) investigated the

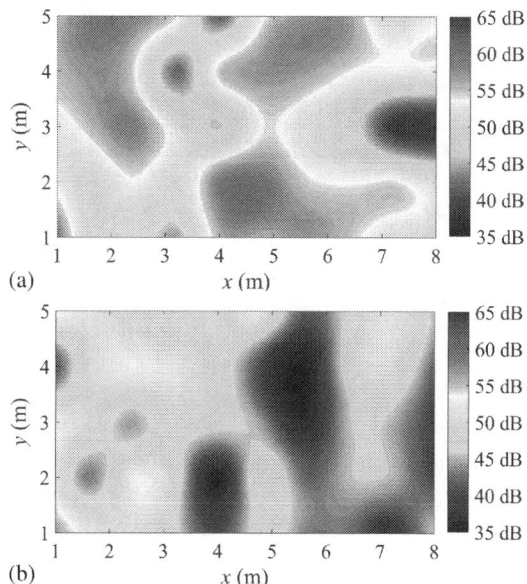

(a)

(b)

**FIGURE 4.2**
Sound pressure distribution at 100 Hz on the evaluation plane (*x–y* plane): (a) without the planar virtual sound barrier, (b) with the planar virtual sound barrier.

effect of the opening area of windows on the sound reduction index and found that partially opened double glazing was 10 dBA better than partially opened single glazing, and that the double glazing could be opened up to 100 mm to reach the noise insulation capacity of the closed single glazing. Later, Field and Fricke (1998) demonstrated that quarter-wave resonators of different lengths could be used outside building ventilation openings to reduce the amount of noise entering buildings over a wide frequency range with an attenuation of about 7 dB in certain 1/3 octave bands.

Kang and Brocklesby (2005) developed a staggered window system, where the opening sashes of a spaced double-glazed window were staggered to create a natural ventilation path and prevent direct sound propagation, and transparent micro-perforated absorbers were used along the ventilation path to attenuate external noise. It was found that external noise could be efficiently reduced in the frequency range from 500 Hz to 8000 Hz; however, the performance of the staggered window system is not satisfactory at low frequencies. Yuya et al. (2009) presented a model for soundproofing casement windows by optimizing the locations of input and output openings as well as the input area size.

Tong and Tang (2013) investigated the insertion losses of plenum windows installed on a building facade for a non-parallel line source with a 1:4 scaled-down model in a semi-anechoic chamber, and found that the insertion losses were around 5–18 dB, depending on different orientation situations. Two years later, a full-scale field measurement was carried out with two identical mock-up test rooms, the same dimensions as those commonly adopted for Hong Kong public housing near a busy trunk road; one was equipped with the plenum windows, and the other with the conventional side-hung casement windows (Tong et al., 2015). Four internal room settings were compared, and the results showed that the acoustical benefit achieved by replacing the side-hung casement windows with the plenum windows tested was around 7.1–9.5 dBA.

The sound insulation performance of passive windows is usually poor in the low-frequency range. Active noise control (ANC) techniques have been applied to the problem. In 2002, a model was proposed for examining the coherence between reference and error signals of traffic noise transmission through an open window into a rectangular room in high-rise buildings (Zhang, Jiang, and Li, 2002). Ise (2005) arranged 16 independent single-channel ANC systems at an open window to reflect the incident noise back and achieved more than 10 dB noise reduction at the error sensors in the frequency range from 200 Hz to 700 Hz. The research progress on natural ventilation ANC windows was summarized and reported by Qiu, Huang, and Lin (2011).

Nishimura and his team proposed the active acoustic shielding (AAS) system, which consists of a number of identical AAS cells in an array (Nishimura, Ohnishi, and Kanamori, 2008). Figure 4.3 shows the basic concept of the AAS for sound transmission control through an open window and a diagram of an AAS cell. Each AAS cell is a simple feedforward active control system with a reference microphone located as close as possible in front of a secondary source (the noise source side). With the nearly collocated reference microphone and secondary source, the controller is expected to reduce incident sound from different directions (Murao et al., 2014). Because of this,

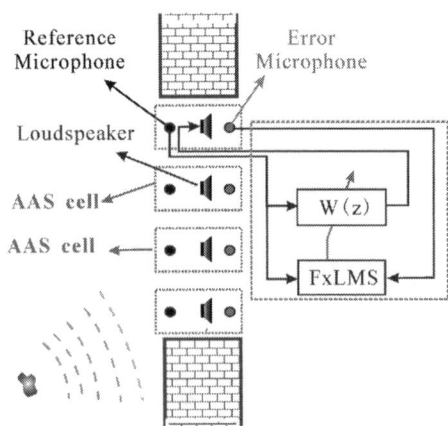

**FIGURE 4.3**
Basic concept of the AAS system and a diagram of the AAS cell.

a howling compensation filter usually has to be adopted to reduce the acoustical feedback from the secondary source to the reference microphone (Murao, Nishimura, and Sakurama, 2012). The control filter $W(z)$ for canceling the sound in front of the cell can be designed off-line or can be updated online in an adaptive system with the filtered-x LMS algorithm (Elliott, 2001).

A small AAS window with a size of $250 \times 250$ mm² was made with four AAS cells and installed in the door of an anechoic room. The measurement results show that the AAS window can effectively reduce the random incident noise transmission in the frequency range from 500 Hz to 2000 Hz (Murao et al., 2014). The challenge is that large numbers of AAS cells are necessary for large windows in practical use, so new decentralized controllers and the related algorithms need to be developed (Murao et al., 2016). The FPGA (field programmable gate array) circuit has been used to develop control hardware to reduce system latency (Shi et al., 2017). A full-scale model with an actual domestic sliding window and security grille was constructed to test the performance of the distributed control system, where 16 channels of secondary sources were evenly distributed across the window opening, and notable attenuation of more than 5 dB (energy average SPL) was achieved (Lam et al., 2018b). Figure 4.4 shows a photo of such a system developed at Nanyang Technological University (Shi et al., 2016).

A mixed-error approach has been proposed to reduce the computational complexity of multichannel active noise control systems for mitigating noise passing through open windows (Murao et al., 2017). In their proposed system, the reference signals are sensed at the front of the secondary loudspeakers, and only one reference signal is used in each feedforward channel to reduce the computational complexity. But a number of error microphones are used to ensure global noise reduction of the system, and the mixed-error approach proposes to add the error signals in advance so that the number of inputs to the controller is reduced. Simulation and experimental results confirm that the performance of global noise reduction is not compromised by the mixed-error approach in a 4-channel system applied to a 0.2 m by 0.2 m

**FIGURE 4.4**
The experimental setup of the open window ANC system developed at Nanyang Technological University (Shi et al., 2016).

open window. For practical applications, it is desired to remove the error microphones. Although many approaches, such as using the open-loop control (Shi et al., 2016), the fixed-filter control (Lam et al., 2016), and the incoming sounds classification control (Ranjan et al., 2016), have been investigated, there is still no mature solution for removing the error sensors in the systems at present.

## 4.3 Implementation Issues

Although virtual sound barrier systems have the potential to be applied to many possible scenarios, there are also many issues for implementing the technology. This section uses the application to open windows as an example in order to discuss these issues.

Many researchers in the world have been contributing to the research on sound transmission control through open windows. Both passive and active approaches can be applied. Passive approaches use sound absorption and reflection materials or structures on the opening or its boundaries to absorb or reflect incident sound, but there are two challenges in practice. One is the compromise between sound transmission control and open access or ventilation, and the other is the heavy mass and large volume requirements of passive approaches in the low-frequency range. Active control approaches can provide a better solution thanks to the superior low-frequency performance with small-size and lightweight structures. The research on the active control of sound via open windows can be roughly divided into two categories (Qiu, 2017).

The first one is the application of the virtual sound barriers, which control sound transmission through window openings directly by installing secondary sources on the openings or at the edges of the openings to reduce sound propagation through the openings. The advantages of this category are their simple installation and high air circulation rate; however, they have poor performance in the middle- to high-frequency range and a potential high cost because of the requirements of using a lot of secondary sources. The second one is the ANC plenum windows, which apply ANC on the plenum structures transformed from opening windows. In the systems in this category, the plenum chambers provide noise attenuation in the middle- to high-frequency range while, at the same time, providing a certain degree of airflow for natural ventilation; also the ANC components attenuate low-frequency components of noise that propagate through the plenum chambers. The plenum chambers transform the original three-dimensional space sound propagation problem into a one-dimensional sound propagation problem in a duct or a low-order acoustic cavity control problem, so that ANC can be implemented more efficiently, with better performance and at lower cost. The disadvantages of this category are their complicated structures and that there can be some air exchange rate decrease.

There is still no mature application of virtual sound barrier systems. Even for the noise reduction windows, there is still no such window that has both good noise insulation and natural ventilation. Many issues need to be addressed to make successful, commercial natural ventilation window products. For example, for the first category systems that use the virtual sound barriers directly, the challenge is that lots of secondary sources are needed for a relatively large window, leading to less feasibility for practical applications. Future research directions can be focused on the development of compact decentralized multiple adaptive ANC controllers, and the understanding of the mechanisms and physics of the systems with random incident sound. Because the secondary sources located in the middle of the opening are sometimes not desired in some applications, applying the secondary sources only at the edges of the openings might be a solution. It has been found that an upper-limit frequency exists for effective control when the secondary sources are installed only at the edges of an opening, so it might be necessary to have lattice-form ANC windows, where a large window is divided into many small lattices, with secondary sources being installed on the frames of these lattices.

For the practical implementation of the virtual sound barrier systems on open windows, the size of the control sound sources needs to be reduced, so thin, small, powerful loudspeaker systems with reasonably good low-frequency responses are required. Furthermore, the robustness and performance of the systems need to be improved so that they can achieve good noise reduction under all circumstances, such as with strong wind, heavy rain, and thunderstorms. Finally, the cost of the controllers and the systems needs to be reduced so that mass production markets and building industries can make use of them. For both categories, short delay AD/DA converters and powerful digital signal processors (DSPs) or FPGAs are necessary due to the causality requirement for the real-time control. Specially designed control chips might be necessary to bring the cost down.

# 5

## Summary and Perspectives

### 5.1 Summary

A virtual sound barrier is an active noise control system that uses arrays of loudspeakers and microphones to create a useful quiet zone in a noisy environment, just like an ordinary sound barrier but without much blocking of air, light, and access. This system can be used to reduce sound propagation, radiation, and transmission from noise sources, or to reduce noise levels around a few persons in a noisy environment.

The first chapter introduces the concepts in sound propagation and the principles of various sound barriers that are used for sound propagation control. The basic wave phenomena related to sound propagation such as acoustic reflection, absorption, scattering, and diffraction are introduced first, and then the corresponding parameters such as reflection coefficient, absorption coefficient, scattering coefficient, transmission loss, and insertion loss are explained. The principles of passive sound barriers and active sound barriers are discussed next, as well as the methods used for their design. Finally, the history, principles, and design methods of virtual sound barriers are introduced.

The second chapter focuses on planar virtual sound barriers. The problem is defined first, so that the concept of planar virtual sound barriers can be introduced. The use of planar virtual sound barriers is then applied to the control of plane wave propagation in free fields, the control of plane wave propagation through a finite aperture, the control of sound radiation from an opening of an enclosure, and the control of sound transmission via an opening into an enclosure. Both the boundary control and surface control systems are discussed, and their upper-limit effective frequencies are considered. The methods for designing planar virtual sound barriers for these applications are given and the mechanisms for the control are explored.

The third chapter introduces three-dimensional virtual sound barriers, which can be used to create a quiet zone in a noisy environment, just like a passive sound barrier, but without much blocking of air, light, and access. The effects of a diffracting sphere inside the quiet zone, a reflective surface near the system, and the cost functions for optimizing the system for better performance are discussed. The virtual sensor algorithms developed for local active noise control systems are reviewed, and the effects of virtual sensor locations on system performance are discussed. It is found that system performance can be improved significantly with the use of virtual sensors when the distances between the secondary sources and the physical sensors are small.

Virtual sound barrier systems can be implemented in situations where both sound control and free air ventilation are required. The fourth chapter introduces two examples of virtual sound barrier applications for noise control. The first is the control of noise radiation by power transformers in a hemi-closed space with a planar virtual sound barrier, where an array of loudspeakers is installed evenly at the open door of the room to reduce the noise radiation generated by the transformers inside. The second is the control of sound transmission into a room through an open window, where an array of loudspeakers is installed at the open window of the room to reduce the transmission of noise generated by outside traffic into the room. Finally, the implementation issues relating to these applications are discussed.

Although the existing research has demonstrated that it is feasible to develop virtual sound barrier systems, there are still obstacles to their practical use, due to the narrow bandwidth and low-frequency range that can be used for effective control, the complexity, and the high cost of the systems. Chapter 5 summarizes the research progress made concerning virtual sound barriers, lists the problems and challenges, and gives the perspectives. If the upper-limit frequency of virtual sound systems can be increased, they have the potential to be applied to many sound control scenarios with ventilation and/or access requirements.

As demonstrated in the previous chapters, the research on virtual sound barrier systems has made significant progress. The related theories and models have been established, most of the mechanisms have been understood, and there are also a few examples of industry applications. But unlike active noise control headphones, which are successful commercial products used in everyday life, there are still no successful mass-produced products based on virtual sound barrier systems. Even for the two example applications described in Chapter 4, only prototypes have been developed so far for testing. There is still a long way to go in order to bring about a wider application of virtual sound barriers.

## 5.2 Perspectives

The existing theoretical studies, numerical simulations, and experiments on virtual sound barrier systems and other related research have demonstrated their feasibility. In a complex sound field, where noise comes from many different directions, a quiet zone of a certain size can be generated using a virtual sound barrier system, especially in the low-frequency range. However, it is still hard to apply these systems in practice, due to the narrow bandwidth and low-frequency range of effective control, the complexity, and the high cost. Along with the increase of the upper-limit frequency of effective noise reduction, the number of secondary sources increases dramatically with the square of the frequency. For example, with the 16-channel virtual sound barrier system described in Section 3.2.5, the upper-limit frequency is about 500 Hz for having an average reduction of more than 10 dB inside a cylindrical space of 0.2 m height and 0.2 m radius. If the upper-limit frequency of the system increases to 4000 Hz, the required number of secondary sources needs to be up to 1024, which is hardly practical.

### 5.2.1 Future Research Topics

It was shown in Section 3.5 that the upper-limit frequency can be increased, or the size of the quiet zone can be enlarged, by using a cost function for minimizing the sum of the total acoustic energy density at the error sensors (Zou, Qiu, and Lu, 2009). To implement this error-sensing strategy, the sound pressure and three orthogonal components of the particle velocities need to be measured by a microphone and three particle velocity sensors respectively. The adaptive active control algorithm for minimizing the sum of the total acoustic energy density needs to be developed, which includes the approaches for secondary path modeling with four error signals (one sound pressure and three particle velocities in three directions) and the methods for reducing the algorithm complexity and memory space requirements by using the redundancy information from these four signals.

Particle velocity transducers are usually more expensive than microphones in practice, so it is preferred to use microphones. The problem with sound pressure-sensing strategies is that effective noise reduction cannot be achieved at certain frequencies due to the interior Dirichlet problem with zero sound pressure at the boundary, as shown in Section 3.5.2. The double-layer sensor array is a potential solution to the problem. Although there are preliminary studies relating to the optimization of the secondary source directivity of virtual sound barrier systems (Poletti, Abhayapala, and Samarasinghe, 2012), experimental research has not been carried out. A general solution with a double-layer sensor array and a double-layer secondary source array, as illustrated in Figure 5.1, can be investigated both theoretically and experimentally (Chang and Jacobsen, 2013).

It is desired in practice that the use of error sensors in virtual sound barrier systems can be omitted. The main difficulty is how to ensure effective sound control in different primary sound fields. A possible solution is using a feedforward control system with fixed control parameters, as shown in Figure 5.2. The reference sensor array is arranged outside the secondary source array, and the control signal of the loudspeakers is determined by

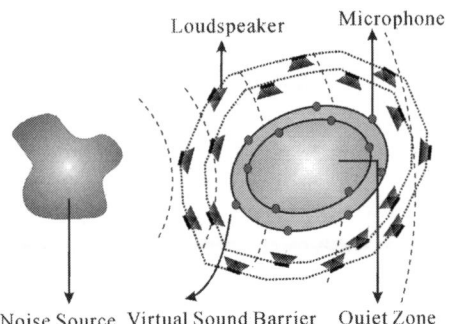

**FIGURE 5.1**
Schematic drawing of a double-layered virtual sound barrier system.

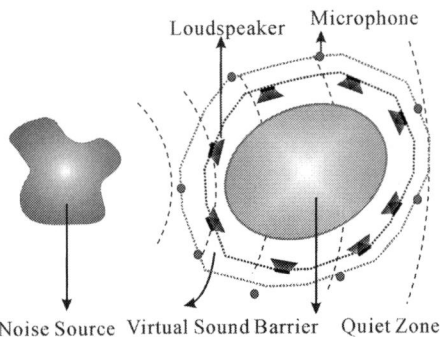

**FIGURE 5.2**
Schematic drawing of a mapping virtual sound barrier system without using error sensors.

mapping equations, which include the relationships between the sound field information on the surface of the reference sensor array, the sound field information inside the virtual sound barrier system, and the relationships between the input of secondary sources and the sound field information inside the system. All these relationships, as well as the robustness of the mapping equations to various primary sound fields, can be investigated in the future.

Hybrid virtual sound barrier systems use different acoustic elements at the quiet zone boundary, where sound field control is obtained by integrating functions of passive elements with different impedance and adaptive active elements. A hybrid active and passive control method might improve the performance of a virtual sound barrier system by increasing the noise reduction inside the quiet zone, or increasing the upper-limit frequency of effective sound control, or reducing the secondary source numbers (Pan and Qiu, 2008). One such kind of hybrid virtual sound barrier system is composed of some active secondary sources and several passive sound insulation partitions, which are arranged at intervals, as shown in Figure 5.3(a).

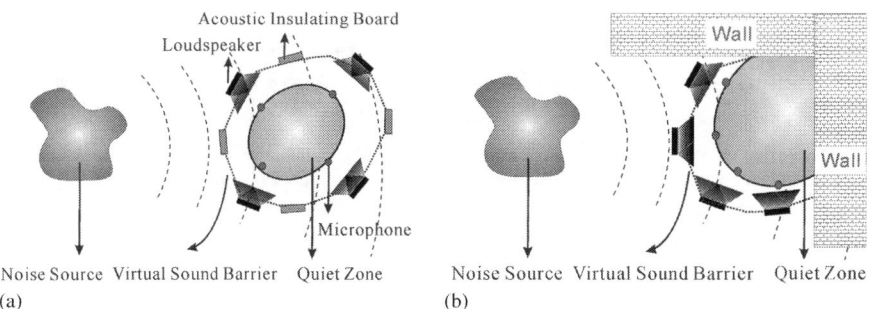

**FIGURE 5.3**

Schematic drawings of: (a) a partition hybrid virtual sound barrier system, (b) a corner hybrid virtual sound barrier system.

The physical mechanisms of this hybrid virtual sound barrier system are not clear, and the effects of the intervals, the size, and the arrangement of the partitions on the performance of the system can be analyzed. Figure 5.3(b) shows another kind of hybrid system, where the virtual sound barrier system is located in or close to a designed acoustic structure, such as a corner. Future studies can be on the physical mechanisms of the systems, the effect of the acoustic characteristics of wall surfaces on control performance, the optimization methods of the acoustic parameters, and the geometry of the structures.

In addition to new architectures for virtual sound barrier systems, novel compound or directional secondary sources and sensors, or virtual sensors, can be developed to increase the performance of the systems. Virtual sound barrier systems are completed active control systems, which adopt loudspeaker and microphone arrays and use signal processing techniques to control the sound field. From the algorithm aspects, there are also many problems that can be explored, such as the stability condition of non-ideal and insufficient length secondary path models and their influence on the performance of adaptive control algorithms. To further overcome the detrimental effect of imperfect secondary path models, the control algorithms without secondary path modeling are of particular interest. To reduce the complexity and cost of practical virtual sound barrier systems, it is also important to investigate the implementation of different feedforward, feedback, and/or integrated control structures, as well as the centralized and decentralized control architectures.

In some virtual sound barrier systems, the transfer functions among the actuators and error sensors have similar characteristics, so a simplified plant estimation technique might be applied by exploiting the spatial-temporal structure of the multiple secondary propagation paths. Using a specific cascaded structure for the multichannel cancellation path impulse responses might reduce the number of parameters for the cancellation path model significantly. A multichannel delay-less sub-band active control algorithm with cascaded cancellation path modeling can be developed, where the cascaded structure for cancellation path modeling can be extended to the sub-band and combined with the multichannel delay-less sub-band active control system. The multi-delay adaptive filter is a flexible structure, which partitions a long filter into many shorter sub-filters, so that a much smaller FFT (fast Fourier transform) size can be used to reduce both the delay and the memory requirements, while maintaining the low-computational complexity and faster convergence properties of the frequency domain algorithms.

## 5.2.2 Challenges for the Applications

Although theories and models of virtual sound barriers have been developed, there are still many challenges for making real products. Regarding planar virtual sound barriers, one expectation from practice is a system that looks like a passive noise control partition or panel. The sensors, the actuators, and the control circuits are all integrated into the planar structure, and the planar virtual sound barrier acts just like a passive noise control barrier for blocking sound propagation, but it has few objects in the propagation path to enable free access and ventilation. Figure 5.4 shows two examples of potential planar virtual sound barrier products. One is a window without glass and the other is a sound reduction panel with many holes. In both systems, the loudspeakers and microphones are installed on the boundaries or frames of the window to allow natural ventilation and free access.

Some of the challenges for transforming the designs to products are listed below.

- Although it is theoretically possible to use a fixed-coefficient controller for such a planar virtual sound barrier system in order to obtain some sound reduction performance, a specific control system (either feedback or feedforward structure) needs to be carefully designed and tested for a non-adaptive system under different practical conditions.

- For an adaptive control system, an error-sensing strategy is needed to use the information from the sensors installed on the boundaries of the openings to estimate the sound power transmitted into the other space or in a few directions to the other side.

- For a feedforward control system, reference sensors have to be used, but the acoustic feedback from the secondary sources to the reference sensors degrades the system performance, or even makes the system unstable, so some kinds of anti-howling algorithms or circuits need to be developed.

- Although a centralized system can achieve the maximum performance, a low-cost decentralized multiple-channel system is desired in practice. The scalability of the system has the advantage of flexibility for openings of different shapes and sizes. Therefore, effective decentralizing algorithms for multiple-channel control systems need to be developed that can achieve similar performances to the centralized ones but maintain the system stability.

- To have a successful commercial product, the controller, the signal-conditioning hardware for the sensors, and the power amplifiers for the loudspeakers should be integrated into a single electronic chip to reduce system cost and to increase system reliability.

- To help with practical applications of the planar virtual sound barrier systems, hardware and software tools and design handbooks need to be developed to help engineers to design different systems for different applications.

Figure 5.5 shows two example systems, which can be developed in the future for three-dimensional virtual sound barriers. They are in the shape of hemisphere to create a quiet zone around a seat or on a bed. The loudspeakers and microphones are on the surface of the hemisphere, and the space between

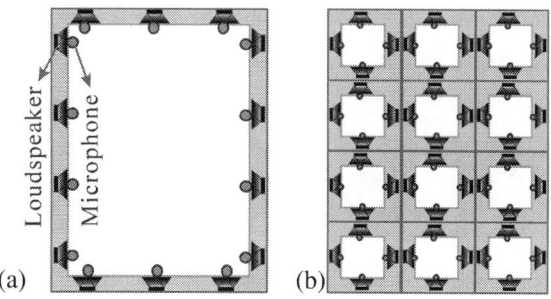

**FIGURE 5.4**

Schematic drawings of planar virtual sound barrier product examples: (a) a window without glass, (b) a sound reduction panel with many holes.

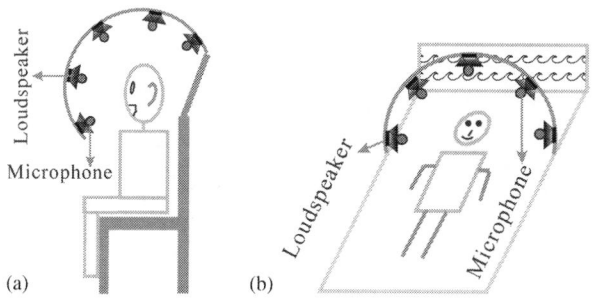

**FIGURE 5.5**

Schematic drawings of three-dimensional virtual sound barrier product examples: (a) a quiet zone generator around a seat, (b) a quiet zone generator on a bed.

the loudspeakers can be empty to let air and light pass through. Most of the challenges for transforming these designs into products are similar to those for the planer virtual sound barriers, but there are some different ones. For example,

- In an enclosure or in a room, the sound comes from many directions due to the reflections from the boundaries, so a method of obtaining good reference signals is needed for a feedforward system to meet the requirements of good coherence as well as causality.
- Because the noise in a room is usually broad band, a hybrid system that incorporates both passive and active components is needed in order to reduce the system cost as well as to improve the overall performance of the system.

# References

Attenborough, K., Li, K., and Horoshenkov, K. (2007). *Predicting Outdoor Sound*. Taylor & Francis, Abingdon, UK.

Berkhoff, A. P. (2005). Control strategies for active noise barriers using near-field error sensing. *Journal of the Acoustical Society of America*, 118:1469–1479.

Bies, D. A., Hansen, C. H., and Howard, C. Q. (2018). *Engineering Noise Control*, 5th Edition. CRC Press, Boca Raton, FL.

Boone, M. M. and Ouweltjes, O. (1997). Design of a loudspeaker system with a low-frequency cardioidlike radiation pattern. *Journal of the Audio Engineering Society*, 45:702–707.

Bowman J. J., Senior T. B. A., and Uslenghi P. L. E. (1969). *Electromagnetic and Acoustic Scattering by Simple Shapes*. North-Holland Publishing, Amsterdam, Netherlands.

Burnett, D. S. and Soroka, W. W. (1972). Tables of rectangular piston radiation impedance functions, with application to sound transmission loss through deep apertures. *Journal of the Acoustical Society of America*, 51:1618–1623.

Chang, J. H. and Jacobsen, F. (2013). Experimental validation of sound field control with a circular double-layer array of loudspeakers. *Journal of the Acoustical Society of America*, 133:2046–2054.

Chen, W., Rao, W., Min, H., and Qiu, X. (2011). An active noise barrier with unidirectional secondary sources. *Applied Acoustics*, 72:969–974.

Chen, W., Min, H., and Qiu, X. (2013). Noise reduction mechanisms of active noise barriers. *Noise Control Engineering Journal*, 61:120–126.

Cox, T. J. and D'Antonio, P. (2009). *Acoustic Absorbers and Diffusers: Theory, Design and Application*, 2nd Edition. CRC Press, London.

Cunefare, K. A. and Shepard, S. (1993). The active control of point acoustic sources in a half-space. *Journal of the Acoustical Society of America*, 93:2732–2739.

de Lacerda, L. A., Wrobel, L. C., Power, H., and Mansur, W. J. (1998). A novel boundary integral formulation for three dimensional analysis of thin acoustic barriers over an impedance plane. *Journal of the Acoustical Society of America*, 104:671–678.

Duhamel, D. (1996). Efficient calculation of the three dimensional sound pressure field around a noise barrier. *Journal of Sound and Vibration*, 197:547–471.

Elliott, S. J. (2001). *Signal Processing for Active Control*. Academic Press, London.

Elliott, S. J. and Cheer, J. (2015). Modelling local active sound control with remote sensors in spatially random pressure fields. *Journal of the Acoustical Society of America*, 137:1936–1946.

Elliott, S. J., Cheer, J., Lam, B., Shi, C., and Gan, W. S. (2018). A wavenumber approach to analysing the active control of plane waves with arrays of secondary sources. *Journal of Sound and Vibration*, 41:405–419.

Epain, N. and Friot, E. (2007). Active control of sound inside a sphere via control of the acoustic pressure at the boundary surface. *Journal of Sound and Vibration*, 299:587–604.

Field, C. D. and Fricke, F. R. (1998). Theory and applications of quarter-wave resonators: A prelude to their use for attenuating noise entering buildings through openings. *Applied Acoustics*, 53:117–132.

Ford, R. D. and Kerry, G. (1973). The sound insulation of partially open double glazing. *Applied Acoustics*, 6:57–72.

Garcia-Bonito, J. and Elliott, S. J. (1995). Local active control of diffracted diffuse sound fields. *Journal of the Acoustical Society of America*, 98:1017–1024.

Garcia-Bonito, J. and Elliott, S. J. (1999). Active cancellation of acoustic pressure and particle velocity in the near field of a source. *Journal of Sound and Vibration*, 221:85–116.

Garcia-Bonito, J., Elliott, S. J., and Bonilha, M. (1997). Active cancellation of pressure at a point in a pure tone diffracted diffuse sound field. *Journal of Sound and Vibration*, 201:43–65.

Garcia-Bonito, J., Elliott, S. J., and Boucher, C. C. (1997). Generation of zones of quiet using a virtual microphone arrangement. *Journal of the Acoustical Society of America*, 101:3498–3516.

Guo, J. and Pan, J. (1997). Further investigation on actively created quiet zones by multiple control sources in free space. *Journal of the Acoustical Society of America*, 102:3050–3053.

Guo, J. and Pan, J. (1998a). Effects of reflective ground on the actively created quiet zones. *Journal of the Acoustical Society of America*, 103:944–952.

Guo, J. and Pan, J. (1998b). Increasing the insertion loss of noise barriers using an active-control system. *Journal of the Acoustical Society of America*, 104:3408–3416.

Guo, J. and Pan, J. (1999). Actively created quiet zones for broadband noise using multiple control sources and error microphones. *Journal of the Acoustical Society of America*, 105:2294–2303.

Guo, J., Pan, J., and Bao, C. (1997). Actively created quiet zones by multiple control sources in free space. *Journal of the Acoustical Society of America*. 101:1492–1501.

Hadden, W. J. and Pierce, A. D. (1981). Sound diffraction around screens and wedges for arbitrary point source locations. *Journal of the Acoustical Society of America*, 69:1266–1276.

Hansen, C. H., Snyder, S. , Qiu, X., Brooks, L., and Moreau, D. (2013). *Active Control of Noise and Vibration*, Second Edition. CRC Press, Boca Raton, FL.

Han, N. and Qiu, X. (2007). A study of sound intensity control for active noise barriers. *Applied Acoustics*, 68:1297–1306.

Hart, C. R. and Lau, S. K. (2012). Active noise control with linear control source and sensor arrays for a noise barrier. *Journal of Sound and Vibration*, 331:15–26.

Hesselmann, N. (1978). Investigation of noise reduction on a 100 kVA transformer tank by means of active methods. *Applied Acoustics*, 11:27–34.

Huang, H., Qiu, X., and Kang, J. (2011). Active noise attenuation in ventilation windows. *Journal of the Acoustical Society of America*, 130:176–188.

Ingard, U. (1951). On the reflection of a spherical sound wave from an infinite plane. *Journal of the Acoustical Society of America*, 23(3):319–335.

Ise, S. (2005). The boundary surface control principle and its applications. *IEICE Transactions on Fundamentals of Electronics, Communications and Computer Sciences*, E88-A:1656–1664.

Isei, T., Embleton, T. F. W., and Piercy, J. E. (1980). Noise reduction by barriers on finite impedance ground. *Journal of the Acoustical Society of America*, 67:46–58.

International Organization for Standardization. (1994). *Acoustics – Determination of sound power levels of noise sources using sound pressure – Engineering method in an essentially free field over a reflecting plane* (ISO Standard 3744:1994). Available at: www.iso.org/standard/9240.html.

Johnson, M. E., Elliott, S. J., Baek, K. H., and Garcia-Bonito, J. (1998). An equivalent source technique for calculating the sound field inside an enclosure containing scattering objects. *Journal of the Acoustical Society of America*, 104:1221–1231.

Jonasson, H. G. (1972). Sound reduction by barriers on the ground. *Journal of Sound and Vibration*, 22:113–126.

Joseph, P., Elliott, S. J., and Nelson, P. A. (1994). Near field zones of quiet. *Journal of Sound and Vibration*, 172:605–627.

Kang, J. and Brocklesby, M. W. (2005). Feasibility of applying micro-perforated absorbers in acoustic window systems. *Applied Acoustics*, 66:669–689.

Keller, J. B. (1962). Geometrical theory of diffraction. *Journal of the Optical Society of America*, 52:116–130.

Kestell, C. D., Cazzolato, B. S., and Hansen, C. H. (2000). Active noise control with virtual sensors in a long narrow duct. *International Journal of Acoustics and Vibration*, 5:63–76.

Kestell, C. D., Cazzolato, B. S., and Hansen, C. H. (2001). Active noise control in a free field with virtual error sensors. *Journal of the Acoustical Society of America*, 109:232–243.

Kinsler, L. E., Frey, A. R., Coppens, A. B., and Sanders, J. V. (2000). *Fundamentals of Acoustics*. John Wiley & Sons, New York.

Kleinman, R. E. and Roach, G. F. (1974). Boundary integral equations for the three-dimensional Helmholtz equation. *SIAM Review*, 16:214–236.

Kurze, U. J. and Anderson, G. S. (1971). Sound attenuation by barriers. *Applied Acoustics*, 4:35–53.

Lau, S. K. and Tang, S. K. (2009). Investigation of system configuration and pressure gradient control for active noise barrier. *Proceedings of the 38th International Congress & Exhibition on Noise Control Engineering* (INTER-NOISE 2009) Ottawa, Canada, II, 1250–1259. Noise Control Foundation/Curran Associates, Red Hook, NY.

Lam, B., Elliott, S., Cheer, J., and Gan, W. S. (2018a). Physical limits on the performance of active noise control through open windows. *Applied Acoustics*, 137:9–17.

Lam, B., He, J., Murao, T., Shi, C., Gan, W. S., and Elliott, S. J. (2016). Feasibility of the full-rank fixed-filter approach in the active control of noise through open windows. *Proceedings of the 45th International Congress & Exhibition on Noise Control Engineering* (INTER-NOISE 2016), Hamburg, Germany. Deutsche Gesellschaft für Akustik. Available at: http://pub.dega-akustik.de/IN2016/data/articles/000488.pdf.

Lam, B., Shi, C., Shi, D., and Gan, W. S. (2018b). Active control of sound through full-sized open windows. *Building and Environment*, 141:16–27.

L'Esperance, A., Nicolas, J., and Daigle, G. A. (1989). Insertion loss of absorbent barriers on ground. *Journal of the Acoustical Society of America*, 86:1060–1064.

Li, K. M. and Wong, H. Y. (2005). A review of commonly used analytical and empirical formulae for predicting sound diffracted by a thin screen. *Applied Acoustics*, 66:45–76.

Lin, Z., Lu, J., Shen, C., Qiu, X., and Xu, B. (2004). Active control of radiation from a piston set in a rigid sphere. *Journal of the Acoustical Society of America*, 115:2954–2963.

Liu, J. C. and Niu, F. (2008). Study on the analogy feedback active soft edge noise barrier. *Applied Acoustics*, 69:728–732.

MacDonald, H. M. (1915). A class of diffraction problem. *Proceedings of the London Mathematical Society*, 14:410–427.

Maekawa, Z. (1968). Noise reduction by screens. *Applied Acoustics*, 1:157–173.

Mangiante, G. (1977). Active sound absorption. *The Journal of the Acoustical Society of America*, 61:1516–1523.

Mangiarotty, R. A. (1963). Acoustic radiation damping of vibrating structures. *The Journal of the Acoustical Society of America*, 35:691–707.

Menounou P. A. (2001). A correction to Maekawa's curve for the insertion loss behind barriers. *Journal of the Acoustical Society of America*, 110:1828–1838.

Min, H. and Qiu, X. (2009). Multiple acoustic diffraction around rigid parallel wide barriers. *Journal of the Acoustical Society of America*, 126:179–186.

Moreau, D., Cazzolato, B., Zander, A., and Petersen, C. (2008). A review of virtual sensing algorithms for active noise control. *Algorithms*, 2008:69–99.

Morfey, C. L. (2001). *Dictionary of Acoustics*. Academic Press, London.

Morgan, P. A., Hothersall, D. C., and Chandler-Wilde, S. N. (1998). Influence of shape and absorbing surface-A numerical study of railway sound barriers. *Journal of Sound and Vibration*, 217:405–417.

Morse, P. M. and Ingard, K. U. (1968). *Theoretical Acoustics*. McGraw-Hill, New York.

Munn, J. M., Cazzolato, B. S., Kestell, C. D., and Hansen, C. H. (2003). Virtual error sensing for active noise control in a one-dimensional waveguide: Performance prediction versus measurement. *Journal of the Acoustical Society of America*, 113:35–38.

Murao, T. and Nishimura, M. (2012). Basic study on active acoustic shielding. *Journal of Environmental Engineering*, 7:76–91.

Murao, T., Nishimura, M., He, J., Lam, B., Ranjan R., Shi, C., and Gan, W. S. (2016). Feasibility study on decentralized control system for active acoustic shielding. *Proceedings of the 45th International Congress and Exposition on Noise Control Engineering* (INTER-NOISE 2016), Hamburg, Germany. Deutsche Gesellschaft für Akustik. Available at: http://pub.dega-akustik.de/IN2016/data/articles/000154.pdf.

Murao, T., Nishimura, M., and Sakurama, K. (2012). Basic study on active acoustic shielding: Phase 4 improving noise reducing performance in low frequency-2. *Proceedings of the 41th International Congress and Exposition on Noise Control Engineering* (INTER-NOISE 2012), New York, USA, II, 1205–1216. Noise Control Foundation/Curran Associates, Red Hook, NY.

Murao, T., Nishimura, M., Sakurama, K., and Nishida, S. (2014). Basic study on active acoustic shielding: Phase 6 improving the method to enlarge AAS window-2. *Proceedings of the 43th International Congress and Exposition on Noise Control Engineering* (INTER-NOISE 2014), Melbourne, Australia. Australian Acoustical Society. Available from: www.acoustics.asn.au/conference_proceedings/INTERNOISE2014/papers/p483.pdf.

Murao, T., Shi, C., Gan, W. S., and Nishimura, M. (2017). Mixed-error approach for multi-channel active noise control of open windows. *Applied Acoustics*, 127:305–315.

Nelson, P. A. and Elliott, S. J. (1992). *Active Control of Sound*. Academic Press, London.

Nishimura, M., Ohnishi, K., and Kanamori, N. (2008). Basic study on active acoustic shielding. *Proceedings of the 37th International Congress and Exposition on Noise Control Engineering* (INTER-NOISE 2008), Shanghai, China, I, 246–261. Institute of Acoustics, Chinese Academy of Sciences/ Curran Associates, Red Hook, NY.

Niu, F., Zou, H., Qiu, X., and Wu, M. (2007). Error sensor location optimization for active soft edge noise barrier. *Journal of Sound and Vibration*, 299:409–417.

Ohnishi, K., Saito, T., Teranishi, S., Namikawa, Y., Mori, T., Kemura, K., and Uesaka, K. (2004). Development of the product-type active soft edge noise barrier. *Proceedings of 18th International Congress on Acoustics*, II, 1277–1280, Kyoto, Japan. Acoustical Society of Japan. Available at: www.icacommission.org/Proceedings/ICA2004Kyoto/pdf/Tu2.F.5.pdf.

Omoto, A. and Fujiwara, K. (1993). A study of an actively controlled noise barrier. *Journal of the Acoustical Society of America*, 94:2173–2180.

Omoto, A., Takashima, K., Fujiwara, K., Aoki, M., and Shimizu, Y. (1997). Active suppression of sound diffracted by a barrier: An outdoor experiment. *Journal of the Acoustical Society of America*, 102:1671–1679.

Pan, J. and Bies, D. (1990). The effect of fluid-structural coupling on sound waves in an enclosure— Theoretical part. *The Journal of the Acoustical Society of America*, 87:691–707.

Pan, J. and Qiu, X. (2008). Performance of an active control system near a reflecting surface. *Australian Journal of Mechanical Engineering*, 5:35–42.

Pathak, P. H., Carluccio, G., and Albani, M. (2013). The uniform geometrical theory of diffraction and some of its applications. *IEEE Antennas and Propagation Magazine*, 55:41–69.

Piacentini, A., Invernizzi, M., and Pannesse, L. (1996). Computational acoustics: Noise reduction via diffraction by barriers with different geometries. *Computer Methods in Applied Mechanics and Engineering* 130:81–91.

Petersen, D., Zander, A. C., Cazzolato, B. S., and Hansen, C. H. (2005). Optimal virtual sensing for active noise control in a rigid-walled acoustic duct. *Journal of the Acoustical Society of America*, 118:3086–3093.

Petersen, C. D., Zander, A. C., Cazzolato, B. S., and Hansen, C. H. (2007). A moving zone of quiet for narrowband noise in a one-dimensional duct using virtual sensing. *Journal of the Acoustical Society of America*, 121:1459–1470.

Pierce, A. D. (1981). *Acoustics*. McGraw-Hill, New York.

Poletti, M. A., Abhayapala, T. D., and Samarasinghe, P. (2012). Interior and exterior sound field control using two dimensional higher-order variable-directivity sources. *The Journal of the Acoustical Society of America*, 131:3814–3823.

Qiu, X. (2017). Recent advances on active control of sound transmission through ventilation windows. *Proceedings of the 24th International Congress on Sound and Vibration* (ICSV24), London, UK. II, 834–841. The International Institute of Acoustics and Vibration/Curran Associates, Red Hook, NY.

Qiu, X., Huang, H., and Lin, Z. (2011). Progress in research on natural ventilation ANC windows. *Proceedings of the 40th International Congress and Exposition on Noise Control Engineering* (INTER-NOISE 2011), Osaka, Japan, III, 2566–2573. Institute of Noise Control Engineering Japan & Acoustical Society of Japan/ Curran Associates, Red Hook, NY.

Qiu, X., Li, N., and Chen, G. (2005). Feasibility study of developing practical virtual sound barrier system. *Proceedings of the 12th International Congress on Sound and Vibration* (ICSV12), Lisbon, Portugal, I, 474–481. The International Institute of Acoustics and Vibration/Curran Associates, Red Hook, NY.

Qiu, X., Zhang L., and Tao J. (2012). Progress in research on active control of transformer noise. *Proceedings of the 41st International Congress & Exhibition on Noise Control Engineering* (INTER-NOISE 2012), New York City, USA. Institute of Noise Control Engineering.

Qiu, X. and Zhao S. (2015). Active control of the directivity of the sound diffraction from barriers. *Proceedings of the 22nd International Congress on Sound and Vibration* (ICSV22), Florence, Italy, II, 929–936. The International Institute of Acoustics and Vibration/Curran Associates, Red Hook, NY.

Qiu, X. and Zou, H. (2016). Recent progress in research on virtual sound barriers. *Proceedings of Acoustics 2016, The Second Australasian Acoustical Societies Conference*, Brisbane, Australia. The Australian Acoustical Society. Available at: https://opus.lib.uts.edu.au/bitstream/10453/108508/1/p163.pdf.

Qiu, X., Zou, H., and Rao, W. (2009). Performance of a virtual sound barrier near a reflective surface. *Proceedings of the 7th International Symposium on Active control of Sound and Vibration* (Active 2009), Ottawa, Canada, 239–249. Noise Control Foundation /Curran Associates, Red Hook, NY.

Rafaely, B., Garcia-Bonito, J., and Elliott, S. J. (1999). Broadband performance of an active headrest. *Journal of the Acoustical Society of America*, 106:787–793.

Ranjan, R., He, J. Murao, T., Lam, B., and Gan, W. S. (2016). Selective active noise control system for open windows using sound classification. *Proceedings of the 45th International Congress and Exposition on Noise Control Engineering* (INTER-NOISE 2016), Hamburg, Germany. Institute of Noise Control Engineering. Available at: http://pub.dega-akustik.de/IN2016/data/articles/000265.pdf.

Rao, W. (2011). A study of effects of boundary and directional sound sources on virtual sound barrier systems. Master Degree Thesis, Nanjing University, Nanjing.

Ross, C. F. (1978). Experiments on the active control of transformer noise. *Journal of Sound and Vibration*, 61:473–480.

Roure, A. and Albarrazin, A. (1999). The remote microphone technique for active noise control. *Proceedings of the 7th International Symposium on Active control of Sound and Vibration* (Active 2009), Ottawa, Canada. II, 1233–1244. Noise Control Foundation /Curran Associates, Red Hook, NY.

Rudnick, I. (1947). The propagation of an acoustic wave along a boundary. *Journal of the Acoustical Society of America*, 19:348–356.

Sgard, F., Nelisse, H., and Atalla, N. (2007). On the modeling of the diffuse field sound transmission loss of finite thickness apertures. *Journal of the Acoustical Society of America*, 122:302–313.

Sha, K., J. Yang, and W. S. Gan. (2005). A simple calculation method for the self-and mutual-radiation impedance of flexible rectangular patches in a rigid infinite baffle. *Journal of Sound and Vibration*, 282:179–195.

Shao, J, Sha, J. Z., and Zhang, Z. L. (1997). The method of the minimum sum of squared acoustic pressures in an actively controlled noise barrier. *Journal of Sound and Vibration*, 204:381–385.

Shepard, W. S. and Cunefare, K. A. (1994). Active control of extended acoustic sources in a half-space. *Journal of the Acoustical Society of America*, 96:2262–2271.

Shi, C., Murao, T., Shi, D., Lam, B., and Gan, W. S. (2016). Open loop active control of noise through open windows. *Proceeding of the 172nd Meeting of the Acoustical Society of America* (Proc. Mtgs. Acoust. 29, 030007), Honolulu, Hawaii. The Acoustical Society of America. Available at: https://asa.scitation.org/doi/pdf/10.1121/2.0000461?class=pdf.

Shi, D., He, J., Shi, C., Murao, T., and Gan, W. S. (2017). Multiple parallel branch with folding architecture for multichannel filtered-x least mean square algorithm. *Proceedings of 2017 IEEE International Conference on Acoustics, Speech and Signal Processing* (ICASSP). Institute of Electrical and Electronics Engineers. Available at: https://ieeexplore.ieee.org/document/7952344.

Sommerfeld, A. (1896). *Mathematical Theory of Diffraction.* Translated by Nagem, R. J., Zampolli, M., and Sandri, G. Progress in Mathematical Physics, 35, 1–68, Birkhäuser Basel, 2004.

Tao, J., Qiu, X., and Pan, J. (2015). Control of transformer noise using an independent planar virtual sound barrier. *Proceedings of ACOUSTICS 2015,* Hunter Valley, Australia. The Australian Acoustical Society. Available at: www.acoustics.asn.au/conference_proceedings/AAS2015/papers/p123.pdf.

Tao, J., Wang, S., and Qiu, X. (2015). Applying virtual sound barrier at a room opening for transformer noise control. *Proceedings of the 22nd International Congress on Sound and Vibration* (ICSV22), Florence, Italy, II, 961–968. The International Institute of Acoustics and Vibration/Curran Associates, Red Hook, NY.

Tao, J., Wang, S., Qiu, X., Xue, J., and Pan, J. (2016). Performance of an independent planar virtual sound barrier at the opening of a rectangular enclosure. *Applied Acoustics*, 105:215–223.

Tao, J., Zou, H., and Qiu, X. (2010). Roof structure design in active noise barrier. *Proceedings of the 17th International Congress on Sound and Vibration* (ICSV17), Cairo, Egypt, I, 355–362. The International Institute of Acoustics and Vibration/Curran Associates, Red Hook, NY.

Thomasson, S. I. (1978). Diffraction by a screen above an impedance boundary. *Journal of the Acoustical Society of America*, 63:1768–1781.

Tong, Y. G. and Tang, S. K. (2013). Plenum window insertion loss in the presence of a line source — A scale model study. *Journal of the Acoustical Society of America*, 133:1458–1467.

Tong, Y. G., Tang, S. K., Kang, J., Fung, A., and Yeung, M. K. L. (2015). Full scale field study of sound transmission across plenum windows. *Applied Acoustics*, 89:244–253.

Wang, S., Tao, J., and Qiu, X. (2015). Performance of a planar virtual sound barrier at the baffled opening of a rectangular cavity. *Journal of the Acoustical Society of America*, 138:2836–2847.

Wang, S., Tao, J., and Qiu, X. (2017). Controlling sound radiation through an opening with secondary loudspeakers along its boundaries. *Scientific Reports*, 7:13385.

Wang, K., Tao, J., and Qiu, X. (2019). Boundary control of sound transmission into a cavity through its opening. *Journal of Sound and Vibration*, 442:350–365.

Wang, S., Tao, J., Qiu, X., and Pan, J. (2018). Mechanisms of active control of sound radiation from an opening with boundary installed secondary sources. *Journal of the Acoustical Society of America*, 143:3345–3351.

Wang, S., Yu, J., Qiu, X., Pawelczykc. M., Shaid, A., and Wang, L. (2017). Active sound radiation control with secondary sources at the edge of the opening. *Applied Acoustics*, 117:173–179.

Williams, E. (1999). *Fourier Acoustics: Sound Radiation and Nearfield Acoustical Holography*. Academic Press, New York.

Wong, H. Y. and Li, K. M. (2001). Prediction models for sound leakage through noise barriers. *Journal of the Acoustical Society of America*, 109:1011–1022.

Yang, C., Pan, J., and Cheng, L. (2013). A mechanism study of sound wave-trapping barriers. *Journal of the Acoustical Society of America*, 134:1960–1969.

Yang, J. and Gan, W. S. (2001). On the actively controlled noise barrier. *Journal of Sound and Vibration*, 240:592–597.

Yuya, N, Sohei, N, Tsuyoshi, N., and Takashi, Y. (2009). Sound propagation in sound-proofing casement windows. *Applied Acoustics*, 70:1160–1167.

Zhang, J., Jiang, W., and Li, N. (2002). Theoretical and experimental investigations on coherence of traffic noise transmission through an open window into a rectangular room in high-rise buildings. *Journal of the Acoustical Society of America*, 112:1482–1495.

Zhang, L., Tao, J., and Qiu, X. (2012). Active control of transformer noise with an internally synthesized reference signal. *Journal of Sound and Vibration*, 331:3466–3475.

Zhao, S., Cheng, E., Qiu, X., Larcy, J., and Maisch, S. (2017). A method of configuring fixed coefficient active noise controllers for traffic noise reduction. *Proceedings of the 46th International Congress and Exposition on Noise Control Engineering* (INTER-NOISE 2017), Hong Kong. Institute of Noise Control Engineering, 3762–3769. International Institute of Noise Control Engineering/Curran Associates, Red Hook, NY.

Zhao, S., Qiu, X., and Cheng, J. (2015). An integral equation method for calculating sound field diffracted by a rigid barrier on an impedance ground. *Journal of the Acoustical Society of America*, 138:1608–1613.

Zou, H. (2007). A study on virtual sound barrier system in frequency domain. PhD Thesis, Nanjing University, Nanjing.

Zou, H., Huang, X., Hu, S., and Qiu, X. (2014). Applying an active noise barrier on a 110 KV power transformer in Hunan. *Proceedings of the 43rd International Congress & Exhibition on Noise Control Engineering* (INTER-NOISE 2014), Melbourne, Australia. Institute of Noise Control Engineering. Available at: www.acoustics. asn.au/conference_proceedings/INTERNOISE2014/papers/p855.pdf.

Zou, H., Lu, J., and Qiu, X. (2010). The active noise barrier with decentralized feedforward control system. *Proceedings of the 17th International Congress on Sound and Vibration* (ICSV17), Cairo, Egypt. The International Institute of Acoustics and Vibration I, 301–308. The International Institute of Acoustics and Vibration/ Curran Associates, Red Hook, NY.

Zou, H. and Qiu, X. (2008). Performance analysis of the virtual sound barrier system with a diffracting sphere. *Applied Acoustics*, 69:875–883.

Zou, H. and Qiu, X. (2009). Optimization of the locations of the virtual sensors in the virtual sound barrier system. *Journal of Nanjing University*, 45:57–66 (in Chinese).

Zou, H., Qiu, X., and Li, N. (2015). A numerical study of virtual sound barrier with virtual sensors. *Proceedings of the 44th International Congress & Exhibition on Noise Control Engineering* (INTER-NOISE 2015), San Francisco California, USA. Institute of Noise Control Engineering. Available at: http://or.nsfc.gov.cn/bitstr eam/00001903-5/509816/1/1000014141106.pdf.

Zou, H., Qiu, X., and Lu, J. (2009). A study of cost functions for the virtual sound barrier system. *Proceedings of 38th International Congress & Exhibition on Noise Control Engineering* (INTER-NOISE 2009), Ottawa, Canada, VII, 4767–4775. Noise Control Foundation/Curran Associates, Red Hook, NY.

Zou, H., Qiu, X., Lu, J., and Niu, F. (2007). A preliminary experimental study on virtual sound barrier system. *Journal of Sound and Vibration*, 307:379–385.

Zou, H., Tao, J., and Qiu, X. (2017). Present status and future development for active control of transformer noise. *Proceedings of 46th International Congress & Exhibition on Noise Control Engineering* (INTER-NOISE 2017), Hong Kong, China, 4408–4414. International Institute of Noise Control Engineering/Curran Associates, Red Hook, NY.

# *Index*